Lecture Notes in Mathematics

Edited by A. Dold and B. Eckmann

428

Algebraic and Geometrical Methods in Topology

Conference on Topological Methods in Algebraic Topology SUNY Binghamton, October 3-7, 1973

Edited by L. F. McAuley

Springer-Verlag
Berlin · Heidelberg · New York 1974

Dr. Louis F. McAuley
Department of Mathematical Sciences
State University of New York at Binghamton
Binghamton, NY 13901/USA

AMS Subject Classifications (1970): 18D99, 28A75, 55B15, 55D99, 55F05, 55F10, 57B99, 57C10, 57D20, 57D35, 57D50, 57E25, 58B05, 58B10

ISBN 3-540-07019-2 Springer-Verlag Berlin · Heidelberg · New York
ISBN 0-387-07019-2 Springer-Verlag New York · Heidelberg · Berlin

Offsetdruck: Julius Beltz, Hemsbach/Bergstr.

INTRODUCTION

This volume consists of the proceedings of the supplementary program of the Five Day Regional Conference on Topological Methods in Algebraic Topology - A History of Classifying Spaces held at the State University of New York at Binghamton, October 3 - 7, 1973. It was the first conference held on this subject in the world.

The conference was supported by a grant from the National Science Foundation. The principle speaker of the conference was Professor E. E. Floyd, Robert C. Taylor Professor of Mathematics, The University of Virginia. His series of ten lectures should appear as a publication of the Conference Board of the Mathematical Sciences (U.S.A.). The supplementary program was an extremely important part of the conference and is represented by the manuscripts herein. The State University of New York provided a grant from its program "Conversations in the Disciplines" which partially supported the supplementary program. We are indebted not only to the National Science Foundation and the State University of New York but also to all who participated in the conference and contributed so much to its success.

The supplementary program covered a wide variety of topics which assisted in making the conference an extremely interesting one. All lectures in this program were given by invitation. These lectures were on topics of current research interest in algebraic, geometric, and differential topology. They have varying degrees of relationship to the central theme. Some attempt has been made to group them by subject as indicated in the table of contents.

Papers in Section I involve various aspects of homotopy theory with the paper of Stasheff directly related to the conference theme.

Section II consists of two papers in category theory as related to algebraic topology. The work represented in Section III concerns a variety of topics all in the area of manifold and differential topology. The papers in Section IV and V concern aspects of geometric topology with infinite dimensional manifolds represented in Section IV and differential geometry represented in Section V.

We deeply regret that it is impossible to reprint the following papers which were presented at our conference and which represented an important part of the supplementary program. They are as follows:

Gluck, Krigelman, and Singer; "The Converse to the Gauss-Bonnet Theorem in PL".

Singer, David; "Preassigning curvature on the Two-Sphere".

These will appear in the Journal of Differential Geometry.

Cohen, Marshall; "A Proof that Simple-Homotopy Equivalent Polyhedra are Stably Homeomorphic".

This paper will appear in the Michigan Mathematical Journal.

Heller, Alex; "Adjoint Functors and Bar Constructions".

This paper will appear in Advances in Mathematics.

We are most grateful to Jeanne Osborne for her assistance in the careful preparation for the conference and for the thorough manner in which she handled administrative details. We are particularly indebted to Althea Benjamin for the superb typing of the manuscripts.

We would like to acknowledge the invaluable editorial assistance rendered by Ross Geoghegan and Patricia McAuley of the Department of Mathematical Sciences, State University of New York at Binghamton, who read many of the manuscripts and provided various other editorial services.

We are no less appreciative of the assistance of Naomi Bar-Yosef, Barbara Lamberg, and Elizabeth Newton.

Finally, we are indebted to Springer-Verlag for publishing these proceedings and, in particular, to Alice Peters for her supervisory role.

Louis F. McAuley
State University of New York at Binghamton

TABLE OF CONTENTS

FIVE DAY REGIONAL CONFERENCE ON TOPOLOGICAL METHODS IN ALGEBRAIC TOPOLOGY - A HISTORY OF CLASSIFYING SPACES

October 3 - 7, 1973

PARTICIPANTS

Douglas Anderson	Syracuse University
Peter Andrews	University of Maryland
James Arnold	University of Wisconsin - Milwaukee
Peter R. Atwood	Hamilton College
Edward Bayno	Montclair State College
James Becker	Purdue University
Victor Belfi	Texas Christian University
Howard T. Bell	Shippensburg State College
Israel Bernstein	Cornell University
Edwin H. Betz	University of Pennsylvania - Philadelphia
Terrance Bisson	Duke University
Charles Cassidy	Laval University
T. A. Chapman	University of Kentucky and Institute for Advanced Study at Princeton
Chao-Kun Cheng	State University College at Potsdam
Philip T. Church	Syracuse University
Vaclav Chvatal	University of Montreal
Marshall Cohen	Cornell University
Robert Connelly	Cornell University
Frank Dangello	Shippensburg State College
Forrest Dristy	State University College at Oswego
Michael N. Dyer	Institute for Advanced Study
Frank Farmer	Arizona State University
Edwin E. Floyd	University of Virginia
Herman Gluck	University of Pennsylvania
Ron Goldman	University of Maryland
John Harper	University of Rochester
Christopher Hee	Eastern Michigan University
Alex Heller	City University of New York
L. S. Hersch	University of Tennessee
Peter Hilton	Case-Western Reserve University
W. Holsztynski	Institute for Advanced Study at Princeton
Vernon Howe	University of Arkansas
Peter Jung	Albright College

Paul Kainen	Case Western Reserve University
Jerry Kaminker	Purdue University
L. Richardson King	Davidson College
Donald Knutson	Fordham University
Thomas Lada	North Carolina State University
Dana Latch	Douglas College
Jerome LeVan	Eastern Kentucky University
Lloyd Lininger	University of Maryland
James L'Reureux	West Chester State College
Pierre Malraison	Carleton College
William McArthur	Shippensburg State College
James McNamara	State University College at Brockport
John Milnor	Institute for Advanced Study at Princeton
Eric Nummela	University of Florida
Stavros Papastavridis	Brandeis University
Robert Piacenza	University of Miami
Everett Pitcher	Lehigh University
Jack Sanders	University of Missouri - Columbia
Victor Sapojnikoff	Haverford College
James Schafer	University of Maryland
Eugene Seelbach	State University College at Brockport
Albert O. Shar	University of New Hampshire
Albert Sheffer, Jr.	Rice University
David Singer	Cornell University
David Smallen	Hamilton College
James Stasheff	Temple University
William Thedford	Virginia Commonwealth University
Graham Toomer	Cornell University
Edward Turner	State University of New York - Albany
Jack Ucci	Syracuse University
Gerald Ungar	University of Cincinnati
Donovan H. Van Osdol	University of New Hampshire
Alphonse T. Vasquez	City University of New York
John Walsh	Institute for Advanced Study at Princeton
Kai Wang	State University of New York - Buffalo
Laura Weiss	State University College at Potsdam
Robert Wells	Pennsylvania State University
James West	Cornell University
Clarence Wilkerson	Carleton University
H. E. Winklenkemper	University of Maryland
Edythe Woodruff	Trenton State College

The following were among the faculty and graduate students of the State University of New York at Binghamton who were participants in the Conference:

David Edwards
Ross Geoghegan
Louie Mahony
Patricia McAuley
Prabir Roy
Alan Coppola
Steve Dibner
Ron Fintushel
Eric Robinson

PARALLEL TRANSPORT AND CLASSIFICATION OF FIBRATIONS

by

James D. Stasheff[1]

The simplest example of parallel transport is the field of (parallel) vertical vectors on $I \times I$:

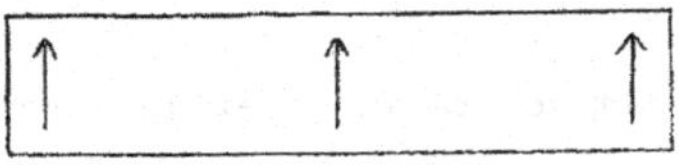

and the simplest non-trivial example occurs when we form this strip into a Moebius band:

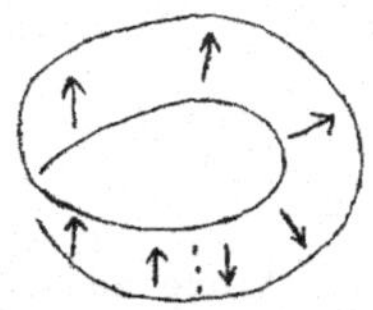

clearly distinguishing the Moebius band from the cylinder.

The idea of parallel transport originates in differential geometry where geometric structure such as curvature is revealed by parallel

[1] Research supported in part by the NSF.

transporting tangent vectors along curves:

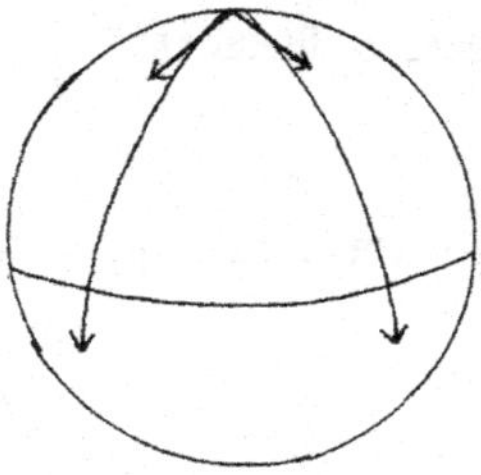

Essentially the same idea occurs in covering space theory where a loop in the space covered determines a deck transformation or permutation of the sheets of the covering. [Veblen and Whitehead] suggested the greater generality of fibre bundles as a setting. We shall look at fiber spaces as well.

We begin formally.

Provisional Definition: For a fibre space $F \longrightarrow E \longrightarrow X$, a (parallel) transport is a map

$$\tau : F \times \Omega X \longrightarrow F$$

such that

1) the trivial loop acts as the identity

2) each loop acts as a homotopy equivalence

3) τ is transitive (i.e., $\tau(f, \lambda + \mu) = \tau(\tau(f,\lambda),\mu)$) or reasonably close to it.

Classically and intuitively we would expect strict transitivity: transporting the fibre around one loop and then another should be the same as transporting it around the sum of the two loops. For fibre spaces we lack such precision as we can see by constructing τ from the Covering Homotopy Property.

Consider

$$\begin{array}{ccc} F \times \Omega X & \xrightarrow{f_o} & E \\ \downarrow & & \downarrow \\ F \times \Omega X & \xrightarrow[g_t]{} & X \end{array}$$

where $f_o(y,\lambda) = y$ and $g_t(y,\lambda) = \lambda(t)$. The CHP gives us $f_t : F \times \Omega X \longrightarrow E$ with $f_1 : F \times \Omega X \longrightarrow F$; in fact, we can assume $f_t(y,e) = y$ where e is the trivial loop. We set $\tau = f_1$ and achieve 1 and 2.

The lifting f_t is not unique, but any two are homotopic. (They are homotopic within E to f_o by a homotopy whose image in X is homotopy trivial and thus the homotopy can be deformed to be fibre preserving, i.e., f_1 and f_1' are homotopic in F.) The same reason applies to show $\tau(\tau \times 1) \simeq \tau \ (1 \times m) : F \times \Omega X \times \Omega X \longrightarrow F$ where m is loop addition [Hilton].

One can in fact say more, but we need a language with which to say it. One approach is to consider the adjoint map ad $\tau : \Omega X \longrightarrow F^F$. (We will not worry about the function space topology but rather always use continuity in reference to τ rather than ad τ). The transitivity of τ is equivalent to the multiplicativity of ad τ. The homotopy condition above is equivalent to ad τ being an H-map. In general for maps of one associative H-space to another we have the notion of strong homotopy multiplicativity.

__Definition.__ If Y and Z are topological monoids, a map $f:Y \longrightarrow Z$ is s.h.m. (strongly homotopy multiplicative) if any of the following conditions are satisfied:

a) There exist maps $f_n : Y^n \times I^{n-1} \longrightarrow Z$ such that $f_1 = f$ and

$$f_n(y_1,\cdots,y_n,t_1,\cdots,t_{n-1}) =$$

$$f_{n-1}(\cdots,y_iy_{i+1},\cdots,\hat{t}_i,\cdots) \quad \text{if} \quad t_i = 0$$

$$f_i(y_1,\cdots,y_i,t_1,\cdots,t_{i-1}) \cdot f_{n-i-1}(y_{i+1},\cdots,y_n,t_{i+1},\cdots,t_{n-1})$$

$$\text{if} \quad t_i = 1 .$$

b) $Sf : SY \longrightarrow SZ$ extends to $BY \longrightarrow BZ$.

c) There exists a commutative diagram

$$\begin{array}{ccc} WY & & \\ \downarrow & \searrow^{h} & \\ Y & \xrightarrow[f]{} & Z \end{array}$$

where $WY \rightarrow Y$ is the standard retraction [Floyd] and h is a homomorphism.

d) f can be factored up to homotopy as $Y \rightarrow Y_1 \leftarrow Y_2 \rightarrow \cdots \rightarrow Z$ where the Y_i are also monoids and the maps are homomorphisms and the maps $Y_{2i} \rightarrow Y_{2i-1}$ are homotopy equivalences.

In particular we can ask if $\mathrm{ad}\ \tau : \Omega X \rightarrow F^F$ is shm. Repeated use of the CHP provides the adjoint maps

4) $$\tau_n : F \times (\Omega X)^n \times I^{n-1} \rightarrow F$$

as desired. Details are given in [10]. The significance of these maps is that they completely determine the fibration as we now indicate.

Let us back up a little. If a group G acts on a space Y, we can look at the orbit space Y/G. If $G \rightarrow Y \rightarrow Y/G$ is not a principal G-bundle, we can replace it, up to homotopy, by one, namely $G \rightarrow EG \times Y \rightarrow EG \times_G Y = Y_G$ where EG is the universal (contractible) G-bundle.

For any fibre space $F \rightarrow E \rightarrow X$, we have the fibration (up to homotopy) $\Omega X \longrightarrow F \longrightarrow E$ which suggests trying to identify E as $F_{\Omega X}$

in some sense. The lack of transitivity is a problem, so let us look at Y_G in more detail. One way of describing Y_G is a realization of the simplicial space

$$\overset{\to}{\underset{\to}{\overset{\to}{\to}}} Y \times G \times G \;\overset{\to}{\underset{\to}{\to}}\; Y \times G \;\overset{\text{action}}{\underset{\text{proj}}{\rightrightarrows}}\; Y \; .$$

In May's notation, the realization is $B(Y,G,*)$, though we have not mentioned degeneracies and prefer to avoid their use, cf. [7].

Now suppose that we have a sh - action of a monoid G on Y (i.e., maps $m_n : Y \times G^n \times I^{n-1} \longrightarrow Y$ adjoint to an shm - map). Form $\coprod_{n \geq 0} Y \times G^n \times I^n$ and factor by the following equivalence relation:

$$(y,g_1,\cdots,g_n,t_1,\cdots,t_n) \sim (\cdots,g_i g_{i+1},\cdots,\hat{t}_i,\cdots) \quad \text{if} \quad t_i = 0$$

$$\sim (m_i(y,\cdots,g_i,t_1,\cdots,t_{i-1})g_{i+1},\cdots,g_n,t_{i+1},\cdots) \quad \text{if} \quad t_i = 1$$

Again call the result Y_G or $B(Y,G,*)$.

In particular all this applies to a transport τ for $F \longrightarrow E \longrightarrow B$.

<u>Theorem:</u> Let $\{\tau_i\}$ be a family of maps satisfying 1), 2), 3) and 4). The map $B(F, \Omega X,*) \longrightarrow B(*, \Omega X,*) = B\Omega X$ is a quasifibration with fibre F. (With extra connective tissue, Fuchs has been able to

build an equivalent Dold fibration [3].)

If τ_i is obtained from $F \to E \to B$ using the CHP as indicated above, then $E \to B$ is weakly fibre homotopy equivalent to $B(F, \Omega X, *) \to B(*, \Omega X, *)$.

If $\{\tau_i\}$ is arbitrary as above and $\{\tau_i'\}$ is constructed from $B(F, \Omega X, *) \to B(*, \Omega X, *)$ using the CHP, then $\{\tau_i\}$ is homotopic to $\{\tau_i'\}$.

Thus $\{\tau_i\}$ is a complete invariant of $E \to B$; homotopy classes of transports classify fibrations.

The usual way of classifying fibrations is by homotopy classes of maps $X \to BH(F)$ where $H(F)$ is the monoid of self-homotopy equivalences of F. Now $\{\mathrm{ad}\, \tau_i\}$ is an shm-map of ΩX into $H(F)$ and hence induces a map at the B level. We have thus

$$X \simeq B\Omega X \xrightarrow{B\mathrm{ad}\tau} BH(F)$$

<u>Theorem.</u> For a suitable choice of the equivalence $X \simeq B\Omega X$, the classifying map above is the usual one [11].

Here we should note that we assume X has the homotopy type of a CW-complex in order to assert $X \simeq B\Omega X$. I am unaware of any study of more general topological conditions (e.g., perhaps weakly locally contractible and paracompact) which would guarantee the same equivalence.

Remarks on operads: Within the context of this conference, it is appropriate to mention the relation between the structures we have been studying and the concept of operads. Our transport $\{\tau_i\}$ is a collection of higher homotopies i.e., maps $F \times (\Omega X)^n \times I^{n-1} \to F$, whereas an operad action is of the form $Y^n \times M(n) \to Y$, where $M(n)$ is a parameter space frequently more complicated than a cube, though often contractible in cases of current interest. An "ancient" example are my complexes K_n e.g., $K_3 = I$ but K_4 is a pentagon

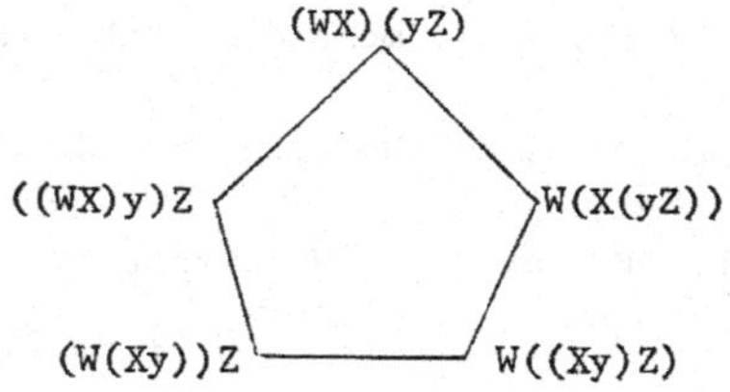

and K_5 a polyhedron with 6 pentagonal and 3 quadrilateral faces. Malraison has a function space equivalent of K_n, readily described in terms of maps $[0,1] \xrightarrow{f} [0,n]$. Thinking of $f^{-1}(i)$ as dividing $[0,1]$ into n pieces, we can see the relevance to loop spaces by using loops parameterized from 0 to 1 and the classical addition of loops. The corresponding K_n structures can be pictured

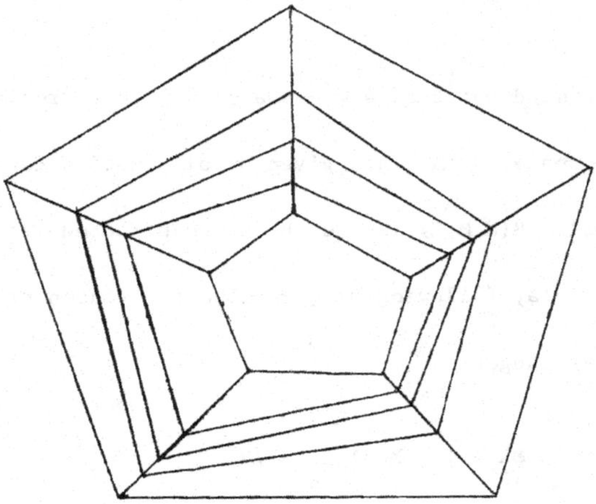

One reason for studying $\{K_n\}$ - spaces rather than strict monoids is that the definition is homotopy invariant. If $X \simeq Y$ and X is a monoid, Y need not admit an equivalent monoid structure (cf. Exotic multiplications on S^3 [Slifker]) but Y will admit an equivalent $\{K_n\}$ - structure (usually called strongly homotopy associative - s.h.a.).

Now recall that an operad is, among other things, a category; where defined, composition is associative. It makes sense to talk of $M \longrightarrow \text{End } X$ being shm rather than a strict morphism, a sh - functor rather than a strict functor. Again if $X \simeq Y$ and X is an M - space, then Y is at least an sh - M - space (Lada). Alternatively Boardman asserts Y is a WM - space where WM is his construction, presented here by Floyd.

Floyd has also pointed out that a WM-space X can be replaced up to homotopy by an M-space. Lada has given an alternative description of this process, namely $B(M,M,X)$ where B is constructed using cubes as above. Actually Lada, following May, usedthe associated triple MX which is just the free gadget

$$MX = \coprod M(n) \times_{\Sigma_n} X^n/\sim$$

where the equivalence is given entirely in terms of degeneracies $d_i : M(n) \longrightarrow M(n-1)$ corresponding to $X^{n-1} \longrightarrow X^n$ by inserting the base point in the i-th coordinate.

In comparing operads by morphisms $M \longrightarrow M^1$ which are homotopy equivalences on each component, we find the inverse maps $M^1 \longrightarrow M$ are at least shm. Finally since operads have associative compositions, we can generalize to sh-operads having operads act on operads.

Since the conference, I have seen work of Segal in which he has related E_∞-Σ-operads to his Γ-structures and given an alternate approach to the last two paragraphs using essentially form d above for handling sh-morphisms.

To come back to more concrete objects, I will consider briefly the "local" approach to classification. Here local refers to structure defined on a space in terms of an open cover $\{U_\alpha\}$. For example, a fibre bundle $p:E \longrightarrow B$ is defined in terms of local product structures:

$$\begin{array}{ccc} p^{-1}(U_\alpha) & \equiv & U_\alpha \times F \\ & \searrow \quad \swarrow & \\ & U_\alpha & \end{array}$$

A fibre space over a nice base [1] can be defined in terms of local equivalences:

$$\begin{array}{ccc} p^{-1}(U_\alpha) & \simeq & U_\alpha \times F \\ & \searrow \quad \swarrow & \\ & U_\alpha & \end{array}$$

A foliation is defined in terms of special local coordinates:

$$U_\alpha \equiv R^k \times R^{n-k}$$

Now an open cover $\{U_\alpha\}$ gives rise to a simplicial space U:

$$\overrightarrow{\overrightarrow{\overrightarrow{\rightarrow}}} \{U_\alpha \cap U_\beta \cap U_\gamma\}_{\alpha,\beta,\gamma} \;\overrightarrow{\overrightarrow{\rightarrow}}\; \{U_\alpha \cap U_\beta\}_{\alpha,\beta} \;\overrightarrow{\rightarrow}\; \{U_\alpha\}$$

where all intersections are non-empty. (If desired, think of $\{U_\alpha\}$ as a category U with $\mathrm{Ob}U = \coprod U_\alpha$, the disjoint union and $\mathrm{Mor}\, U$ given by $\mathrm{Mor}\,(x \in U_\alpha, y \in U_\beta) = \emptyset$ unless $x = y$ in which case $\mathrm{Mor}\,(x,x) = x$). There is an obvious map $BU = |U| \longrightarrow X$ which if X is paracompact is a homotopy equivalence. (The pictures in [9] are quite indicative.)

Now local structures of the sorts considered above imply compatability on the overlaps.

For example, a fibre bundle involves transition functions $g_{\alpha\beta}\colon U_\alpha \cap U_\beta \longrightarrow G$ (where G is the group of the bundle) such that on $U_\alpha \cap U_\beta \cap U_\gamma$, we have

$$g_{\alpha\beta} g_{\beta\gamma} = g_{\alpha\gamma}$$

This is a morphism $U \longrightarrow G$ and hence induces $X \simeq BU \longrightarrow BG$. Classification can be verified directly if we choose the appropriate realization, namely Milnor's which has built in a nice "universal" open cover.

To be precise, for a category C, consider the subset of

$$BC \subset \Delta^\infty \times C^\infty$$

consisting of pairs $(\vec{t}, \{g_{ij}\})$ s/t $\vec{t} \in \Delta^\infty$ and i,j runs over all pairs such that $t_i t_j \neq 0$, $g_{ij} \in \operatorname{Mor} C$ except $g_{ii} \in \operatorname{Ob} C$ and if $t_i t_j t_k \neq 0$ then $g_{ij} g_{jk} = g_{ik}$. Topologize this space by the limit of the quotient topologies of the maps $\Delta^n \times C^{[n]} \longrightarrow BC$ where $C^{[n]}$ denotes composable n-tuples and the map is given by

$$(s_0, \cdots, s_n, g_1, \cdots, g_n) \longrightarrow (\vec{t}, \{g_{ij}\})$$

where $t_{k_j} = s_j$ for some $k_0 < k_1 < \cdots < k_n$ and $g_{k_i k_j} = g_{i+1} \cdots g_j$.

The universal cover of BU is given by $\{t_i^{-1}(0,1]\}$ and the g_{ij} coordinates regarded as functions $U_i \cap U_j \longrightarrow \mathrm{Mor}\, C$ are universal transition functions [9]. (Strictly speaking, the U_i are only point-finite, but following [1] or [6] we can deform the original t_i to functions $\bar{t}_i$ which are locally finite, so the associated $\bar{t}_i^{-1}(0,1]$ are also.)

We now describe the classification procedure. If X is paracompact, we can now restrict attention to countable locally finite coverings $\{U_i\}$. The "1-cocycle" condition $g_{ij}g_{jk} = g_{ik}$ says that $x \mapsto \{g_{ij}(x)\}$ induces a map $BU \longrightarrow BC$. Conversely given a map $X \xrightarrow{f} BC$, define a local structure on X in terms of the covering $\{f^{-1}(U_i)\}$ by $\gamma_{ij}(x) = g_{ij} \circ f\,(x)$ for $x \in U_i \cap U_j$.

Starting from any $f: X \longrightarrow BC$, we obtain $X \longrightarrow B\{f^{-1}U\} \longrightarrow BC$. If we use $t_i \circ f$ as the partition of unity subordinate to $\{f^{-1}U_i\}$, the composite is given by

$$x \longmapsto (\cdots, t_i \circ f(x), \cdots, g_{ij} \circ f(x))$$

but this is precisely how one would represent f in terms of coordinates

t_i and g_{ij}. In the other direction, if we start with a cocycle γ_{ij} on a numerable covering U with associated partition of unity λ_i, then

$$X \longrightarrow BU \longrightarrow BC$$

is given by

$$x \longrightarrow (\lambda_i(x), \gamma_{ij}(x))$$

and this pulls back the universal example to the open cover $\lambda_i^{-1}(0,1] \subset U_i$ with transition functions $\gamma_{ij}(x)$.

Since the same bundle gives rise to different 1-cocycles as we vary the cover or choice of local coordinates, we must also consider equivalence classes of bundles. Following Steenrod [12], two fibre bundles $E_i \longrightarrow X$ for $i = 1,2$ are equivalent if the union of the corresponding families of transition functions can be extended to a 1-cocycle on the union of the corresponding coverings. i.e., if $BU \longrightarrow BC$ and $BV \longrightarrow BC$ extend to $B(U,V)$. The corresponding partitions of unity give rise to maps $\lambda : X \longrightarrow BU \longrightarrow B(U,V)$ and $\mu : X \longrightarrow BV \longrightarrow B(U,V)$ These are homotopic via $t\lambda + (1-t)\mu$ where this really means the obvious linear homotopy in terms of the Δ^∞ coordinates. Thus the classifying maps $X \longrightarrow BC$ for equivalent transition functions are homotopic.

Now for bundles, a bundle $E \longrightarrow X \times I$ is equivalent to $E_0 \times I \longrightarrow X \times I$ where $E_0 = E|X \times 0$; hence homotopic maps induce equivalent bundles. Thus we have the result:

Equivalence classes of G-bundles over X are in 1-1 correspondence with homotopy classes of maps $X \longrightarrow BG$.

In general, the notion of equivalence must be weakened so as to insure that a structure on $X \times I$ implies the equivalence of the restrictions to $X \times t$. This is the approach which works for foliations [4].

For fibre spaces, we have one additional subtlety; we have $g_{\alpha\beta}g_{\beta\gamma}$ only homotopic to $g_{\alpha\gamma}$ in $H(F)$. As discovered by Wirth [14], a specific choice of homotopy

$$g_{\alpha\beta\gamma} : U_\alpha \cap U_\beta \cap U_\gamma \times I \longrightarrow H(F)$$

is crucial to the classification as are higher homotopies

$$U_{\alpha_0} \cap \cdots \cap U_{\alpha_n} \times I^{n-1} \longrightarrow H(F).$$

In other words, we have an shm-map $U \longrightarrow H(F)$ and hence a classifying map

$$X \simeq BU \longrightarrow BH(F)$$

for paracompact X.

Thus whether through the local or the global (e.g., CHP and transport) approach, we see that classification of fibre spaces involves shm-maps. Once again, we can return to strict morphisms by enlarging the operative objects, e.g., $W\Omega X$ or WU, but it is the shm-maps which are the immediate consequences of the defining properties of fibre spaces.

BIBLIOGRAPHY

1. A. Dold, Partitions of unity in the theory of fibrations, Ann. of Math. (2) 78 (1963), 223-255. MR 27 #5264.

2. E. Floyd, this conference.

3. M. Fuchs, A modified Dold-Lashof construction that does classify H-principle fibrations (to appear).

4. A Haefliger, Homotopy and Integrability, Lecture Notes in Mathematics 197.

5. P. Hilton, Homotopy theory and duality, Gordon and Breach, New York, 1965 MR 33 #6624.

6. J. W. Milnor, Construction of universal bundles. II, Ann. of Math. (2) 63 (1956), 430-436.

7. G. Segal, Categories and Cohomology theories.

8. J. D. Stasheff, "Parallel" transport in fibre spaces, Bol. Soc. Mat. Mexicana (2) 11 (1966), 68-84 MR 38 #5219.

9. J. D. Stasheff, Appendices to Bott's lectures on Foliations, Lecture Notes in Math 279.

10. N. E. Steenrod, The classification of sphere bundles, Ann. of Math (2) 45 (1944), 294-311.

11. O. Veblen and J. H. C. Whitehead, The foundations of differential Geometry, Cambridge University Press, 1932.

12. J. F. Wirth, Fibre spaces and the higher homotopy cocycle relations, Thesis, Notre Dame, Ind., 1965.

Localization of nilpotent spaces

by

Peter Hilton*

1. Introduction

The technique of localization was first introduced into topology by Sullivan [11], though it was implicit in Zabrodsky's method of mixing homotopy types [12]. Subsequently it has been exploited by many topologists, e.g., [1,5,7,8,10]. The author, Mislin and Roitberg [5] have used the technique extensively in studying non-cancellation phenomena. A comprehensive treatment of a more general process, executed in the semi-simplicial category, is given in [1]. We give two examples to show the potential of the method.

Example 1.1 Let $V = V_{n+1,2}$ be the Stiefel manifold of unit tangent vectors to S^n. Then V fibres over S^n with fibre S^{n-1} and it follows from a classical theorem of James and Whitehead that V admits a cellular decomposition.

$$V = S^{n-1} \cup e^n \cup e^{2n-1}. \tag{1.1}$$

Moreover, if n is even, then the first attaching map in (1.1) has degree 2 (the Euler characteristic of S^n). Now, as will transpire in Section 2, we may localize cellularly. Thus if we localize at the odd primes, P, then $S^{n-1} \cup e^n$, in (1.1), becomes contractible (since 2 is invertible in the ring of integers localized at P), so that we obtain, from (1.1),

$$V_P \simeq S_P^{2n-1}. \tag{1.2}$$

---

*This is a report on joint work with Guido Mislin and Joseph Roitberg. An expanded version, under joint authorship, will appear as a monograph in the series *Notas de Matematica*.

Of course, the implications of (1.2) for the cohomology of V were already well known. However, (1.2) also enables us to conclude that, for any space Y which can be P-localized, the set of homotopy classes of maps of V into Y_P satisfies

(1.3) $$[V,Y_P] \cong \pi_{2n-1}(Y)_P.$$

Thus, in particular, $[V,Y_P]$ has an abelian group structure. In general, of course, $[X,Y]$ is merely a set with distinguished element, and thus very difficult to handle.

One may say that the traditional tactic in algebraic topology has been to apply an algebraic functor (e.g., homology, cohomology) and then localize at some prime. By localizing *first*, we may gain structure, as in this example.

Example 1.2 Let $S^3 \to E \to S^7$ represent a principal S^3-bundle over S^7. Such a bundle is classified by an element $\alpha \in \pi_6(S^3)$. Now $\pi_6(S^3) = \mathbb{Z}/12$, generated by ω, the Blakers-Massey element which expresses the non-commutativity of the group operation (quaternionic multiplication) on S^3. We will write E_k for the total space E of the bundle classified by $\alpha = k\omega$, $0 \le k \le 11$.. Of course E_k is diffeomorphic to E_ℓ if $k + \ell \equiv 0 \bmod 12$. However, we may prove, by a cellular approximation argument, that $E_k \not\simeq E_\ell$ unless $k \equiv \pm\ell \bmod 12$. For, by the James-Whitehead theorem cited above E_k admits a cellular decomposition

$$E_k = S^3 \cup_{k\omega} e^7 \cup e^{10}.$$

Thus if $E_k \simeq E_\ell$ then $S^3 \cup_{k\omega} e^7 \simeq S^3 \cup_{\ell\omega} e^7$, and from this we rapidly deduce (using a classical desuspension theorem) a commutative square

$$\begin{array}{ccc} S^6 & \xrightarrow{k\omega} & S^3 \\ \downarrow{\scriptstyle \pm 1} & & \downarrow{\scriptstyle \pm 1} \\ S^6 & \xrightarrow{\ell\omega} & S^3 \end{array}$$

It follows that $k\ddot{\omega} = \pm\ell\omega$, so that $k \equiv \pm\ell \bmod 12$.

However, it is easy to prove that, for all primes p,

(1.4)
$$(E_1)_p \simeq (E_7)_p .$$

For we first observe that it is only necessary to localize at the primes $p = 2, 3$, since $\mathbb{Z}/12$ localizes to zero at other primes. Now if ω_p is the localization at p of ω, then ω_2 is of order 4 and ω_3 is of order 3. Thus

$$\omega_2 = -7\omega_2,$$
$$\omega_3 = 7\omega_3,$$

from which (1.4) may be deduced. Indeed it turns out that not only the total spaces E_1, E_7, but also the bundles themselves become (fibre-) homotopy equivalent on localizing at any prime.

The result (1.4) takes on a special interest when one observes that E_1 is the symplectic Lie group $Sp(2)$. E_7 is thus a manifold homotopically distinct from $Sp(2)$, but equivalent to $Sp(2)$ on localizing at any prime. We may also prove [6,8,9] that

(1.5)
$$Sp(2) \times S^3 = E_7 \times S^3, \text{ and}$$

(1.6)
$$Sp(2) \times Sp(2) \simeq E_7 \times E_7 .$$

Either of these relations shows that E_7 is a Hopf manifold. Stasheff, using Zabrodsky's methods, showed [10] that E_7 has the homotopy type of a topological group G. However, we know that E_7 is not a Lie group.

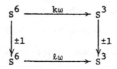

Nor indeed is G; thus G must be infinite-dimensional, since, were it finite-dimensional, it would have to be a manifold and therefore, according to the solution to Hilbert's Fifth Problem, it would admit the structure of a Lie group.

It is interesting to observe in this example that we obtain, by localization techniques, results (like (1.5)) which make no mention of localization. A further such result is, then, that the homotopy analog of Hilbert's Fifth Problem has a negative solution.

In this paper we will be concerned exclusively to construct the localization and to prove the most basic theorem giving the equivalent homotopy and homology characterizations. We will first do this in the homotopy category of 1-connected CW-complexes and will then proceed to generalize to the homotopy category of nilpotent CW-complexes. The generalization will be preceded by a section giving the basic definition and properties of nilpotent spaces. The reader only interested in the 1-connected case should find the section devoted to that case quite self-contained apart from the definition of localization of abelian groups. For the basic notions of localization of abelian and nilpotent groups, the reader is referred to [2,3].

2. Localization of 1-connected CW-complexes

We work in the pointed homotopy category H_1 of 1-connected CW-complexes. if $X \in H_1$, and if P is a family of primes, we say that X is *P-local* if the homotopy groups of X are all P-local abelian groups. We say that $f: X \to Y$ in H_1 *P-localizes* X if Y is P-local and*

$$f^*: [Y,Z] \cong [X,Z]$$

for all P-local $Z \in H_1$. Of course this universal property of f characterizes it up to canonical equivalence: if $f_i: X \to Y_i$, $i = 1, 2$, both P-localize X then there exists a unique equivalence $h: Y_1 \simeq Y_2$ in H_1 with $hf_1 = f_2$. We will prove

Theorem 2A

Every X *in* H_1 *admits a P-localization.*

Theorem 2B

Let $f: X \to Y$ *in* H_1. *Then the following statements are equivalent:*

(i) f *P-localizes* X;

(ii) $\pi_n f: \pi_n X \to \pi_n Y$ *P-localizes for all* $n \geq 1$;

(iii) $H_n f: H_n X \to H_n Y$ *P-localizes for all* $n \geq 1$.

We will prove Theorems 2A, 2B simultaneously. We recall that a homomorphism $\phi: A \to B$ of abelian groups P-localizes if and only if B is P-local and ϕ is a P-isomorphism [2]; this latter conditions means that the kernel and cokernel of ϕ belong to the Serre class C of abelian torsion groups with torsion prime to P. Thus to prove that (ii) $\Leftrightarrow$ (iii) in Theorem 2B above it suffices to prove the following two propositions.

*We write, as usual $[Y,Z]$ for $H_1(Y,Z)$, the set of pointed homotopy classes of maps from Y to Z.

Proposition 2.1

Let $Y \in H_1$. *Then* $\pi_n Y$ *is* P-*local for all* $n \geq 1$ *if and only if* $H_n Y$ *is* P-*local for all* $n \geq 1$.

Proposition 2.2

Let $f: X \to Y$ *in* H_1. *Then* $\pi_n(f)$ *is a* P-*isomorphism for all* $n \geq 1$ *if and only if* $H_n(f)$ *is a* P-*isomorphism for all* $n \geq 1$.

Proof of 2.1 We first observe that $H_n Y$ is P-local for all $n \geq 1$ if and only if $H_n(Y; \mathbb{Z}/p) = 0$ for all $n \geq 1$ and all primes p disjoint from P. Now it was shown in [2] that if A is a P-local abelian group, so are the homology groups $H_n A$, $n \geq 1$. It now follows by induction on m, that if A is a P-local abelian group, so are the homology groups $H_n(A,m)$ of the Eilenberg-MacLane space $K(A,m)$. For we have a fibration $K(A,m-1) \to E \to K(A,m)$, with E contractible, from which we deduce that, if $H_n(A,m-1; \mathbb{Z}/p) = 0$ for all $n \geq 1$, then $H_n(A,m; \mathbb{Z}/p) = 0$ for all $n \geq 1$.

Now let $\ldots \to Y_m \to Y_{m-1} \to \ldots \to Y_2$ be the Postnikov decomposition of Y. Thus there is a fibration $K(\pi_m Y,m) \to Y_m \to Y_{m-1}$, and $Y_2 = K(\pi_2 Y,2)$. Thus, if we assume that $\pi_n Y$ is P-local for all $n \geq 1$, we may assume indúctively that the homology groups of Y_{m-1} are P-local and we infer (again using homology with coefficients in $\mathbb{Z}/p$, with p disjoint from P) that the homology groups of Y_m are P-local. Since $Y \to Y_m$ is m-connected, it follows that $H_n Y$ is P-local for all $n \geq 1$.

To obtain the opposite implication, we construct the 'dual' Cartan-Whitehead decomposition

$$\ldots \to Y(m) \to Y(m-1) \to \ldots \to Y(2).$$

There is then a fibration $K(\pi_m Y,m-1) \to Y(m+1) \to Y(m)$ and $Y(2) = Y$.

Thus, if we assume that H_nY is P-local for all $n \geq 1$, we may assume inductively that the homology groups of $Y(m)$ are P-local. Since $\pi_m Y(m) = \pi_m Y$ and $Y(m)$ is $(m-1)$-connected, it follows that $\pi_m Y \cong H_m Y(m)$ and is P-local. Thus we infer (again using homology with coefficients in $\mathbb{Z}/p$, with p disjoint from P) that the homology groups of $Y(m+1)$ are P-local, so that the inductive step is complete and $\pi_n Y \cong H_n Y(n)$ is P-local.

Proof of 2.2 Since a P-isomorphism is an isomorphism mod C, where C is the class of abelian torsion groups with torsion prime to P, Proposition 2.2 is merely a special case of the classical Serre theorem.

We have those proved that (ii) $\Leftrightarrow$ (iii) in Theorem 2B. We now prove that (ii) $\Leftrightarrow$ (i). The obstructions to the existence and uniqueness of a counterimage of $g: X \to Z$ under $f^*: [Y,Z] \to [X,Z]$ lie in $H^*(f;\pi_* Z)$. Now, given (ii) (or (iii)), $H_* f \in C$. Thus (i) follows from the universal coefficient theorem for cohomology and the following purely algebraic proposition [2].

Proposition 2.3

Let C *be as in the proof of Proposition 2.2. Then, if* $A \in C$ *and* B *is* P-*local,*

$$\mathrm{Hom}(A,B) = 0, \quad \mathrm{Ext}(A,B) = 0.$$

We now prove Thoerem 2A. More specifically, we prove the existence of $f: X \to Y$ in H_1 satisfying (iii). Since we know that (iii) $\Rightarrow$ (i), this will prove Theorem 2A. Our argument is facilitated by the following key observation.

<u>Proposition 2.4</u>

Let U *be a full subcategory of* H_1, *for whose objects* X *we have constructed* $f: X \to Y$ *satisfying (iii). Then the assignment* $X \mapsto Y$ *automatically yields a functor* $L: U \to H_1$, *for which* f *provides a natural transformation from the embedding* $U \subseteq H_1$ *to* L.

<u>Proof of 2.4</u> Let $g: X \to X'$ in U. We thus have a diagram

(2.1)
$$\begin{array}{ccc} X & \xrightarrow{g} & X' \\ \downarrow{\scriptstyle f} & & \downarrow{\scriptstyle f'} \\ Y & & Y' \end{array}$$

in H_1 with f, f' satisfying (iii). Since f satisfies (i) and Y' is P-local by Proposition 2.1, we obtain a unique (in H_1) $h \in [Y,Y']$ making the diagram

(2.2)
$$\begin{array}{ccc} X & \xrightarrow{g} & X' \\ {\scriptstyle f}\downarrow & & \downarrow{\scriptstyle f'} \\ Y & \xrightarrow{h} & Y' \end{array}$$

commutative. It is now plain that the assignment $X \mapsto Y$, $g \mapsto h$ yields the desired functor L.

We exploit Proposition 2.4 to prove, by induction on n, that we may localize all n-dimensional CW-complexes in H_1. If $n = 2$, then such a complex is merely a wedge of 2-spheres,

$$X = \bigvee_{\alpha} S^2,$$

where α runs through some index set, and we define

$$Y = \bigvee_{\alpha} M(\mathbb{Z}_P, 2),$$

where $M(A,2)$ is the Moore space having $H_2M = A$. There is then an evident map $f: X \to Y$ satisfying (iii). Suppose now that we have constructed $f_o: X_o \to Y_o$ satisfying (iii) if $\dim X_o \leq n$, where $n \geq 2$, and let $\dim X = n + 1$, $X \in H_1$. Then we have a cofibration

$$\vee S^n \xrightarrow{g} X^n \xrightarrow{i} X \tag{2.3}$$

By the inductive hypothesis and Proposition 2.4, we may embed (2.3) in the diagram

$$\begin{array}{ccccc} \vee S^n & \xrightarrow{g} & X^n & \xrightarrow{i} & X \\ \downarrow{\scriptstyle f_1} & & \downarrow{\scriptstyle f_o} & & \\ \vee M(\mathbb{Z}_p, n) & \xrightarrow{h} & Y_o & \xrightarrow{j} & Y \end{array} \tag{2.4}$$

where f_o, f_1 satisfy (iii) and the square in (2.4) homotopy-commutes. Thus if j embeds Y_o in the mapping cone Y of h, then we may complete (2.4) by $f: X \to Y$ to a homotopy-commutative diagram and it is then easy to prove (using the exactness of the localization of abelian groups) that $f: X \to Y$ also satisfies (iii). Thus we may construct $f: X \to Y$ satisfying (iii) if X is $(n+1)$-dimensional, and the inductive step is complete.

It remains to construct $f: X \to Y$ satisfying (iii) if X is infinite-dimensional. We have the inclusions

$$X^2 \subseteq X^3 \subseteq \dots \subseteq X^n \subseteq X^{n+1} \subseteq \dots$$

and may therefore construct

$$\begin{array}{ccc} X^n & \overset{i}{\subseteq} & X^{n+1} \\ \downarrow{\scriptstyle f^n} & & \downarrow{\scriptstyle f^{n+1}} \\ Y^{(n)} & \subseteq & Y^{(n+1)} \end{array} \tag{2.5}$$

where f^n, f^{n+1} satisfy (iii). We may even arrange that (2.5) is strictly commutative for each n. If we define $Y = \bigcup_n Y^{(n)}$, with the weak topology, then $Y \in H_1$ and the maps f^n combine to yield a map $f: X \to Y$ which again obviously satisfies (iii). Thus we have proved Theorem 2A in the strong form that, to each X in H_1, there exists $f: X \to Y$ in H_1 satisfying (iii).

Finally, we complete the proof of Theorem 2B by showing that (i) ⇒ (iii). Given $f: X \to Y$ which P-localizes X, let $f_o: X \to Y_o$ be constructed to satisfy (iii). Then $f_o: X \to Y_o$ also satisfies (i), from which one immediately deduces the existence of a homotopy equivalence $u: Y_o \to Y$ with $uf_o \simeq f$. It immediately follows that f also satisfies (iii).

Thus the proofs of Theorems 2A, 2B are complete.

3. Nilpotent spaces

It turns out that the category H_1 is not adequate for the full exploitation of localization techniques. This is due principally to the fact that it does not respect function spaces. We know, following Milnor, that if X is a (pointed) CW-complex and W a finite (pointed) CW-complex, then the function space X^W of pointed maps $W \to X$ has the homotopy type of a CW-complex. However its components will, of course, fail to be 1-connected even if X is 1-connected. However, it turns out that the components of X^W are nilpotent if X is nilpotent. Moreover, the category of nilpotent CW-complexes is suitable for homotopy theory (as first pointed out by E. Dror), and for localization techniques [11].

Definition 3.1 Let G be a group and let A be a G-module. Then we define the *lower central G-series* of A by

$$\Gamma^1 A = A; \quad \Gamma^{n+1}A = \{a - xa,\ a \in \Gamma^n A,\ x \in G\},\ n \geq 1.$$

Moreover, A is said to be G-*nilpotent*, with *nilpotency class* c, where $c \geq 0$, if $\Gamma^c A \neq (0)$, $\Gamma^{c+1}A = (0)$. We also say that G *operates nilpotently* on A if A is G-nilpotent.

Definition 3.2 A connected CW-complex X is *nilpotent* if $\pi_1 X$ is nilpotent and operates nilpotently on $\pi_n X$ for every $n \geq 2$.

Let N be the homotopy category of nilpotent CW-complexes. Plainly $N \supseteq H_1$. Moreover, the *simple* CW-complexes are plainly in N; in particular, N contains all connected Hopf spaces. We prove the following basic theorem.

Theorem 3.3

Let $F \xrightarrow{i} E \xrightarrow{f} B$ *be a fibration of connected* CW-*complexes. Then* $F \in N$ *if* $E \in N$.

Proof We exploit the classical result that the homotopy sequence of the fibration is a sequence of $\pi_1 E$-modules. We will prove that, if $\pi_n E$ is $\pi_1 E$-nilpotent of class $\leq c$, then $\pi_n F$ is $\pi_1 F$-nilpotent of class $\leq c+1$. (A mild modification of the argument is needed to prove that if $\pi_1 E$ is nilpotent of class $\leq c$, then $\pi_1 F$ is nilpotent of class $\leq c+1$; we will deal explicitly with the case $n \geq 2$.)

We will need the fact that $\pi_1 E$ operates on $\pi_n B$ through f_*, and that the operation of $\pi_1 E$ on $\pi_n F$ is such that

(3.1) $$\xi \cdot \alpha = (i_* \xi) \cdot \alpha,\ \xi \in \pi_1 F,\ \alpha \in \pi_n F.$$

It will also be convenient to write I_F, I_E for the augmentation ideals of $\pi_1 F$, $\pi_1 E$. Then the statement that $\pi_n E$ is $\pi_1 E$-nilpotent of class $\leq c$ translates into

(3.2) $$I_E^c \cdot \pi_n E = (0).$$

Consider the exact sequence of $\pi_1 E$-modules

$$\cdots \longrightarrow \pi_{n+1} B \xrightarrow{\partial} \pi_n F \xrightarrow{i_*} \pi_n E \longrightarrow \cdots$$

and let $\xi \in I_F^c$, $\alpha \in \pi_n F$. Then $i_*(\xi \cdot \alpha) = (i_* \xi) \cdot i_*(\alpha) = 0$ by (3.2). Thus $\xi \cdot \alpha = \partial \beta$, $\beta \in \pi_{n+1} B$. Let $\eta \in \pi_1 F$. Then $\partial((i_* \eta - 1) \cdot \beta) = (i_* \eta - 1) . \partial \beta = (i_* \eta - 1) \xi \cdot \alpha, = (\eta - 1) \xi \cdot \alpha$, by (3.1). But $(i_* \eta - 1) \cdot \beta = (f_* i_* \eta - 1) \cdot \beta = 0$, so $(\eta - 1) \xi \cdot \alpha = 0$. This shows that $I_F^{c+1} \cdot \pi_n F = (0)$, and thus the theorem is proved.

Now let W be a finite connected CW-complex and let X be a connected CW-complex. Let X^W be the function space of *pointed* maps $W \to X$ and let X_{fr}^W be the function space of free maps. Choose a map $g \in X^W$ $(g \in X_{fr}^W)$ as base point and let $(X^W, g)((X_{fr}^W, g))$ be the component of g.

Theorem 3.4

(i) (X^W, g) *is nilpotent.*

(ii) (X_{fr}^W, g) *is nilpotent if* X *is nilpotent.*

Proof We may suppose that W^o is a point. Thus the assertions (i), (ii) are certainly true if W is 0-dimensional, and we will argue by induction on the dimension of W. We will be content to prove (i). We have a cofibration

$$V \to W^n \to W^{n+1},$$

where V is a wedge of n-spheres, giving rise to a fibration

$$(X^{W^{n+1}},g) \to (X^{W^n},g_o) \to (X^V,o),$$

where $g: W^{n+1} \to X$ and $g_o = g|W^n$. Our inductive hypothesis is that (X^{W^n},g_o) is nilpotent, so that Theorem 3.3 establishes the inductive step.

<u>Corollary 3.5</u>

Let W *be a finite* CW-*complex and* $X \in N$. *Then* (X^W,g) *and* (X^W_{fr},g) *are nilpotent.*

<u>Proof</u> Let $W_o, W_1, \ldots, W_d$ be the components of W, with $o \in W_o$. Then

$$X^W = X^{W_o} \times X_{f_r}^{W_1} \times \ldots \times X_{fr}^{W_d}.$$

Since plainly a finite product of nilpotent spaces is nilpotent, it follows that (X^W,g) is nilpotent. Similarly (X^X_{fr},g) is nilpotent.

Corollary 3.5 thus establishes (in view of Milnor's theorem) that we stay inside the category N when we take function spaces X^W with $X \in N$ and W finite.

We now proceed to give an important characterization of nilpotent spaces. Let X be a connected CW-complex and let

$$\ldots \longrightarrow X_n \xrightarrow{p_n} X_{n-1} \longrightarrow \ldots \longrightarrow X_1 \longrightarrow o \tag{3.3}$$

be its Postnikov decomposition, so that p_n is a fibration with fibre $K(\pi_n X,n)$. We say that the Postnikov decomposition admits a *principal refinement at stage* n if p_n may be factored as a product of fibrations

(3.4) $$X_n = Y_c \xrightarrow{q_c} Y_{c-1} \longrightarrow \dots \longrightarrow Y_1 \xrightarrow{q_1} Y_o = X_{n-1},$$

where the fibre of q_i is an Eilenberg-MacLane space $K(G_i,n)$ and q_i is induced by a map $g_i\colon Y_{i-1} \to K(G_i,n+1)$, $1 \le i \le c$.

Theorem 3.6

Let X *be a connected* CW*-complex. Then the Postnikov decomposition of* X *admits a principal refinement at stage* $n \ge 2$ *(stage* 1*) if and only if* $\pi_1 X$ *operates nilpotently on* $\pi_n X$ *(*$\pi_1 X$ *is nilpotent).*

Proof We will be content to give the argument for $n \ge 2$. Suppose first that we have the principal refinement (3.4). Then we may regard

$$Y_i \xrightarrow{q_i} Y_{i-1} \xrightarrow{g_i} K(G_i,n+1),\ i = 1, \dots, c.$$

as a fibration. Since $\pi_n Y_o = (0)$, $\pi_1 X$ $(=\pi_1 Y_i, 1 \le i \le c)$ operates trivially on $\pi_n Y_o$. Thus, by repeated applications of the proof of Theorem 3.3, $\pi_1 X$ operates nilpotently on $\pi_n Y_c = \pi_n X_n = \pi_n X$.

Suppose conversely that $\pi_n X$ is $\pi_1 X$-nilpotent of class $\le c$, and that we have factored $p_n\colon X_n \to X_{n-1}$ as

$$X_n \xrightarrow{r_i} Y_i \xrightarrow{s_i} X_{n-1},$$

where $r_i = q_{i+1} \cdots q_c$, each $q_j\colon Y_j \to Y_{j-1}$ being induced by $g_j\colon Y_{j-1} \to K(G_j,n+1)$, $G_j = \Gamma^j \pi_n X/\Gamma^{j+1}\pi_n X$, $i + 1 \le j \le c$. If $i = c$, then $r_c = 1$, $s_c = p_n$. We suppose, moreover, that $q_{j*}\colon \pi_k Y_j \to \pi_k Y_{j-1}$ is the identity for $k < n$, that $q_{j*}\colon \pi_n Y_j \to \pi_n Y_{j-1}$ projects $\pi_n X/\Gamma^{j+1}\pi_n X$ onto $\pi_n X/\Gamma^j \pi_n X$ and that (by consequence) the homotopy groups of each Y_j vanish in dimensions $\ge n + 1$.

We attach (n+1)-cells to Y_i to kill the subgroup $\Gamma^i\pi_n X/\Gamma^{i+1}\pi_n X$ of $\pi_n Y_i = \pi_n X/\Gamma^{i+1}\pi_n X$; let Z' be the resulting space. We then attach (n+2)-cells, (n+3)-cells, ..., to kill $\pi_{n+1}, \pi_{n+2}, \ldots$, so that we have finally embedded Y_i in a space Z, such that the effect on the homotopy groups occurs only in dimension n, where the induced homomorphism is the projection $\pi_n X/\Gamma^{i+1}\pi_n X \twoheadrightarrow \pi_n X/\Gamma^i\pi_n X$. Replace the inclusion $Y_i \subseteq Z$ by a fibration $q_i: Y_i \to Y_{i-1}$. Then the fibre of q_i is $K(G_i,n)$, where $G_i = \Gamma^i\pi_n X/\Gamma^{i+1}\pi_n X$, so that $\pi_1 X = \pi_1 Y_{i-1}$ operates trivially on G_i. It then follows of course that $\pi_1 X$ operates trivially on the homology of $K(G_i,n)$, so that q_i is induced by $g_i: Y_{i-1} \to K(G_i,n+1)$--we take the (negative) transgression of the fundamental class in the fibre. Moreover it is clear that the map q_i is n-connected, so that an easy obstruction argument shows that s_i factors, up to homotopy, uniquely, as $s_i = s_{i-1}q_i$,

$$Y_i \xrightarrow{q_i} Y_{i-1} \xrightarrow{s_{i-1}} X_{n-1}.$$

Thus we continue until we have factored p_n as $s_o q_1 \ldots q_c$, $s_o: Y_o \to X_{n-1}$, with all the fibre maps q_i induced. However, it is now obvious that s_o is a homotopy equivalence, so that we have proved that the Postnikov decomposition of X admits a principal refinement at stage n.

We would say that the Postnikov system of X *admits a principal refinement* if it admits a principal refinement at stage n for every $n \geq 1$. We then have the evident

Corollary 3.7

Let X *be a connected* CW-*complex. Then* X *is nilpotent if and only if its Postnikov system admits a principal refinement.*

We point out that the simple spaces are identified, by the correspondence implicit in this corollary, with those spaces whose Postnikov system is itself principal.

4. Localization of nilpotent complexes

In this section we extend Theorems 2A and 2B from the category H_1 to the category N. To do so we need, of course, to have the notion of the localization of nilpotent groups. This notion, together with the relevant properties, is to be found in [2,3], but we repeat the definition here for the reader's convenience. It will readily be seen that we are generalizing the localization of abelian groups in a very natural way.

Definition 4.1 A nilpotent group G is P-*local* if the function $x \mapsto x^p$, $x \in G$, is bijective for all primes p disjoint from P. A homomorphism $\phi: G \to K$ of nilpotent groups P-*localizes* if K is P-local and

$$\phi^*: \mathrm{Hom}(K,L) \to \mathrm{Hom}(G,L)$$

is bijective for all P-local nilpotent groups L.

Theorem 4.2

Every nilpotent group admits a P-*localization.*

Definition 4.3 Let $X \in N$. Then X is P-*local* if $\pi_n X$ is P-local for all $n \geq 1$. A map $f: X \to Y$ in N P-*localizes* if Y is P-local and

$$f^*: [Y,Z] \cong [X,Z]$$

for all P-local Z in N.

We now come to the main theorems of the paper.

Theorem 4A

Every X *in* N *admits a* P-*localization.*

Theorem 4B

Let $f: X \to Y$ *in* N. *Then the following statements are equivalent:*

(i) f P-*localizes* X*;*

(ii) $\pi_n f: \pi_n X \to \pi_n Y$ P-*localizes for all* $n \geq 1$*;*

(iii) $H_n f: H_n X \to H_n Y$ P-*localizes for all* $n \geq 1$.

The pattern of proof of these theorems will closely resemble that of Theorems 2A, 2B. However, an important difference is that the construction of a localization does not proceed cellularly, as in the 1-connected case, but via a principal refinement of the Postnikov system.

We first prove that (ii) ⇒ (iii) in Theorem 4B. We need a series of lemmas.

Lemma 4.4

If π *acts nilpotently on* A*, then* π *acts nilpotently on* $H_n(A,m)$, $n \geq 0$.

Proof Let $0 = \Gamma^{c+1}A \subseteq \Gamma^c A \subseteq \ldots \subseteq \Gamma^1 A = A$ be the *lower central* π*-series* of A (see Section 3), and write $A_i = \Gamma^i A$ for convenience. Note that each A_i is a nilpotent π-module, of class less than that of A if $i \geq 2$. Moreover, π acts trivially on A_i/A_{i+1}. We have a spectral sequence of π-modules,

$$E_2^{pq} = H_p(A_i/A_{i+1},m;\ H_q(A_{i+1},m)),$$

converging (finitely) to the graded group associated with $H_n(A_i,m)$, suitably filtered. If we assume inductively that π operates nilpotently on $H_q(A_{i+1},m)$, it operates nilpotently on E_2^{pq}, whence it readily follows that π operates nilpotently on $H_n(A_i,m)$, completing the inductive step.

Lemma 4.5

Let $X \in N$ *and let* $\pi = \pi_1 X$. *Then* π *operates nilpotently on* $H_n(\tilde{X})$ *where* $\tilde{X}$ *is the universal cover of* X.

Proof Consider the Postnikov system of $\tilde{X}$. We have a fibration $K(\pi_m X,m) \to \tilde{X}_m \to \tilde{X}_{m-1}$, $m \geq 2$, where $\tilde{X}_1 = o$. Thus we may suppose inductively that π operates nilpotently on the homology of $\tilde{X}_{m-1}$ and, by Lemma 4.4, π operates nilpotently on the homology of $K(\pi_m X,m)$. We have a spectral sequence of π-modules,

$$E_2^{pq} = H_p(\tilde{X}_{m-1};\ H_q(\pi_m X,m)),$$

converging (finitely) to the graded group associated with $H_n\tilde{X}_m$, suitably filtered. We see immediately that π operates nilpotently on E_2^{pq}, whence it readily follows that π operates nilpotently on $H_n\tilde{X}_m$. This completes the inductive step. Since $\tilde{X} \to \tilde{X}_m$ is m-connected, the conclusion of the lemma follows.

Now, if G is a nilpotent group operating on A, and if the action is nilpotent, then, as shown in [3], there is a well-defined induced operation of G_P, the P-localization of G, on A_P, the P-localization of A. Moreover, for these actions,

$$(\Gamma^i A)_P = \Gamma^i A_P.$$

<u>Lemma 4.6</u>

The induced map $H_n(G;A) \to H_n(G_p;A_p)$ *is localization,* $n \geq 0$.

<u>Proof</u>

We argue by induction on the nilpotency class of A. If G operates trivially on A, then this follows from the universal coefficient theorem in homology and the basic result in [2] that

(4.1) $$H_nG \to H_nG_p \text{ localizes, } n \geq 1.$$

We now consider the commutative diagram

$$\begin{array}{ccccc} \Gamma^{i+1}A & \rightarrowtail & \Gamma^iA & \twoheadrightarrow & \Gamma^iA/\Gamma^{i+1}A \\ \downarrow & & \downarrow & & \downarrow \\ \Gamma^{i+1}A_p & \rightarrowtail & \Gamma^iA_p & \twoheadrightarrow & \Gamma^iA_p/\Gamma^{i+1}A_p \end{array}$$

where the vertical arrows are localization. This induces the commutative diagram

$$\begin{array}{ccccccccccc} \cdots H_{n+1}(G;\Gamma^iA/\Gamma^{i+1}A) & \to & H_n(G;\Gamma^{i+1}A) & \to & H_n(G;\Gamma^iA) & \to & H_n(G;\Gamma^iA/\Gamma^{i+1}A) & \to & H_{n-1}(G;\Gamma^{i+1}A) & \to \cdots \\ \downarrow e_1 & & \downarrow e_2 & & \downarrow \hat{e}_3 & & \downarrow e_4 & & \downarrow e_5 & \\ \cdots H_{n+1}(G_p;\Gamma^iA_p/\Gamma^{i+1}A_p) & \to & H_n(G_p;\Gamma^{i+1}A_p) & \to & H_n(G_p;\Gamma^iA_p) & \to & H_n(G_p;\Gamma^iA_p/\Gamma^{i+1}A_p) & \to & H_{n-1}(G_p;\Gamma^{i+1}A_p) & \to \cdots \end{array}$$

Here we know that e_1, e_4 localize and we may assume inductively that e_2, $\hat{e}_5$ localize. It thus follows that e_3 localizes. This completes the inductive step and establishes the lemma.

We are now ready to prove that (ii) ⇒ (iii) in Theorem 4B. Let $\tilde{X}$, $\tilde{Y}$ be the universal covers of X, Y so that we have a diagram

(4.2) $$\begin{array}{ccccc} \tilde{X} & \to & X & \to & K(\pi_1X,1) \\ \downarrow \tilde{f} & & \downarrow f & & \downarrow f_1 \\ \tilde{Y} & \to & Y & \to & K(\pi_1Y,1) \end{array}$$

Since $\tilde{f}$ induces localization in homotopy, it induces localization in homology by Theorem 2B. Moreover, we obtain from (4.2) a map of spectral sequences which is, at the E_2-level,

$$H_p(\pi_1 X; H_q \tilde{X}) \to H_p(\pi_1 Y; H_q \tilde{Y}). \tag{4.3}$$

By Lemma 4.5 $\pi_1 X$ operates nilpotently $H_q \tilde{X}$ and $\pi_1 Y$ operates nilpotently on $H_q \tilde{Y}$. We thus infer from Lemma 4.6, together with (4.1) if $q = 0$, that (4.3) is localization unless $p = q = 0$. Passing through the spectral sequences and the appropriate filtrations of $H_n \tilde{X}$, $H_n \tilde{Y}$, we infer that $H_n f$ localizes if $n \geq 1$.

Now let (t) be the statement $f^*: [Y,Z] \cong [X,Z]$ *for all* P-*local* Z *in* N. Note that this statement differs from (i) only in not requiring that Y be P-local. We prove that (iii) ⇒ (i'). This will, of course, imply that (ii) ⇒ (i).

If Z is P-local nilpotent, then we may find a principal refinement of its Postnikov system. Moreover this principal refinement will be such that the fibre at each stage is a space $K(A,n)$, where A is P-local abelian. For, as we saw in the proof of Theorem 3.6, $A = \Gamma^i \pi_n Z / \Gamma^{i+1} \pi_n Z$ for some i, and it is easy to see [3] that $\Gamma^i B$ is P-local if B is P-local. Given $g: X \to Z$, the obstructions to the existence and uniqueness of a counterimage to g under f^* will thus lie in the groups $H^*(f;A)$ and, as in the corresponding argument in the 1-connected case (note that we have trivial coefficients here, too), these groups will vanish if f induces P-localization in homology.

Next we proceed to prove Theorem 4A, via a key observation playing the role of Proposition 2.4.

<u>Proposition 4.7</u>

Let U *be a full subcategory of* N, *for whose objects* X *we have constructed* $f: X \to Y$ *satisfying (ii). Then the assignment* $X \mapsto Y$ *automatically yields a functor* $L: U \to N$, *for which* f *provides a natural transformation from the embedding* $U \subseteq N$ *to* L.

<u>Proof of 4.7</u> Let $g: X \to X'$ in U. We thus have a diagram

(4.4)
$$\begin{array}{ccc} X & \xrightarrow{g} & X' \\ \downarrow{\scriptstyle f} & & \downarrow{\scriptstyle f'} \\ Y & & Y' \end{array}$$

in N, where f, f' satisfy (ii). Then f satisfies (i) and Y' is P-local, so that there exists a unique h in N such that the diagram

(4.5)
$$\begin{array}{ccc} X & \xrightarrow{g} & X' \\ \downarrow{\scriptstyle f} & & \downarrow{\scriptstyle f'} \\ Y & \xrightarrow{h} & Y' \end{array}$$

commutes. It is now plain that the assignment $X \mapsto Y$, $g \mapsto h$ yields the desired functor L.

We now exploit Proposition 4.7 to prove Theorem 4A. We first consider spaces X in N yielding a *finite* refined principal Postnikov system and, for those, we argue by induction on the *height* of the system. Thus we may assume that we have a principal (induced) fibration

(4.6) $$K(G,n) \to X \to X',$$

where G is abelian even if $n = 1$, and we may suppose that we have constructed $f': X' \to Y'$ satisfying (ii). (The induction starts with $X' = o$.)

Since (4.6) is induced, we may, in fact, assume a fibration

$$X \longrightarrow X' \xrightarrow{g} K(G,n+1)$$

Now we may certainly localize $K(G,n+1)$; we obtain $K(G_P,n+1)$, where G_P is the localization of G and so, by Proposition 4.7, we have a diagram

$$\begin{array}{ccccc} X & \longrightarrow & X' & \xrightarrow{g} & K(G,n+1) \\ & & \downarrow{\scriptstyle f'} & & \downarrow{\scriptstyle e} \\ & & Y' & \xrightarrow{h} & K(G_P,n+1) \end{array}$$

Let Y be the fibre of h. There is then a map $f: X \to Y$ rendering the diagram

$$\begin{array}{ccccc} X & \longrightarrow & X' & \xrightarrow{g} & K(G,n+1) \\ \downarrow{\scriptstyle f} & & \downarrow{\scriptstyle f'} & & \downarrow{\scriptstyle e} \\ Y & \longrightarrow & Y' & \xrightarrow{h} & K(G_P,n+1) \end{array}$$

commutative in N and a straightforward application of the exact homotopy sequence shows that f satisfies (ii).

It remains to consider the case in which the refined principal Postnikov system of X has infinite height (this is, of course the 'general' case!). Thus we have principal fibrations

(4.8) $$\cdots \longrightarrow X_i \xrightarrow{g_i} X_{i-1} \longrightarrow \cdots \longrightarrow o$$

and there is a weak homotopy equivalence $X \to \varprojlim X_i$.

Now we may apply the reasoning already given to embed (4.8) in the diagram, commutative in N,

(4.9)
$$\begin{array}{ccccccc} \cdots \to & X_i & \xrightarrow{g_i} & X_{i-1} & \to \cdots \to & o \\ & \downarrow{\scriptstyle f_i} & & \downarrow{\scriptstyle f_{i-1}} & & \\ \cdots \to & Y_i & \xrightarrow{h_i} & Y_{i-1} & \to \cdots \to & o \end{array}$$

where each f_i satisfies (ii). Moreover, we may suppose that each h_i is a fibre map. Let Y be the geometric realization of the singular complex of $\varprojlim Y_i$. Then there is a map $f: X \to Y$ such that the diagram

$$\begin{array}{ccc} X & \longrightarrow & \varprojlim X_i \\ \downarrow{\scriptstyle f} & & \downarrow{\scriptstyle \varprojlim f_i} \\ Y & \longrightarrow & \varprojlim Y_i \end{array}$$

is homotopy-commutative. Moreover, the construction of (4.9) shows that the Y_i-sequence is again a refined principal Postnikov system, from which it readily follows that $\varprojlim f_i$ satisfies (ii). So therefore does f, and f is in N. Thus we have completed the proof of Theorem 4A in the stronger form that there exists, for each X in N, a map $f: X \to Y$ in N satisfying (ii).

The proof that (i) ⇒ (ii) proceeds exactly as in the easier case of the category H_1. Thus we have established the following set of implications:

(4.10) (ii) ⇒ (iii), (iii) ⇒ (i'), (ii) ⇒ (i), (i) ⇒ (ii).

All that remains is to prove the following proposition, for then we will be able to infer that, in fact, (iii) ⇒ (i).

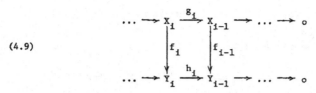

Proposition 4.8

If $Y \in N$ *and* $H_n Y$ *is P-local for every* $n \geq 1$, *then* $\pi_n Y$ *is P-local for every* $n \geq 1$.

To prove this, we invoke Dror's theorem, which we, in fact, reprove since it follows immediately from (4.10). Thus we consider the special case $P = \Pi$, where Π is the collection of all primes. Then a homomorphism of (nilpotent, abelian) groups Π-localizes if and only if it is an isomorphism. Moreover, every space in N is Π-local, so that, in this special case, the distinction between (i') and (i) disappears. Thus (4.10) implies, in particular, the equivalence of (ii) and (iii) for $P = \Pi$, which is Dror's theorem.

Now we prove Proposition 4.8. We construct $f: Y \to Z$ satisfying (ii). It thus also satisfies (iii); but $H_n Y$ is P-local, so that f induces an isomorphism in homology. By Dror's theorem, f induces an isomorphism in homotopy. However, the homotopy of Z is P-local, so that Proposition 4.8 is proved, and, with it, the proof of Theorems 4A and 4B is complete.

Remarks 1. Of course, we do not need the elaborate machinery assembled in this section to prove Dror's theorem. In particular, Theorem 4A is banal for $P = \Pi$, since, then, the identity $X \to X$ Π-localizes!

2. The fact that we have both the homotopy criterion (ii) and the homology criterion (iii) for the localizing map f enables us to derive some immediate conclusions. For example we may use (ii) to prove [4]

Theorem 4.9

If X *is nilpotent and* W *finite and if* $f: X \to Y$ *localizes, then* $f^W: X^W \to Y^W$ *localizes.*

Similarly, we use (ii) to prove

<u>Theorem 4.10</u>

If $F \to E \to B$ *is a fibre sequence in* N, *then so is* $F_P \to E_P \to B_P$, *where* X_P *is the* P-*localization of* X.

Finally, we use (iii) to prove

<u>Theorem 4.11</u>

If $U \to V \to W$ *is a cofibre sequence in* N, *then so is* $U_P \to V_P \to W_P$.

3. An important reason for the difference between the proofs of Theorem 2B and Theorem 4B is that, in H_1, we can construct the localization *cellularly*, whereas in N we construct it *homotopically*. It would be very interesting to know whether the localization can be constructed cellularly in N.

References

1. A. K. Bousfield and D. M. Kan, Homotopy limits, completions and localizations. Lecture Notes in Mathematics 304, Springer (1972).

2. P. J. Hilton, Localization and cohomology of nilpotent groups, Math. Zeit. (1973) (to appear).

3. P. J. Hilton, Remarks on the localization of nilpotent groups, Comm. Pure and Applied Math. (1973) (to appear).

4. P. J. Hilton, G. Mislin and J. Roitberg, Homotopical localization, Proc. Lond. Math. Soc. 3, XXVI (1973), 693-706.

5. P. J. Hilton, G. Mislin and J. Roitberg, H-spaces of rank 2 and non-cancellation phenomena, Inv. Math. 16 (1972), 325-334.

6. P. J. Hilton and Joe Roitberg, On principal S^3-bundles over sphere, Ann. of Math. 90 (1969), 91-107.

7. M. Mimura, G. Nishida and H. Toda, Localization of CW-complexes and its applications, J. Math. Soc. Japan, 23 (1971), 593-624.

8. G. Mislin, The genus of an H-space, Lecture Notes in Mathematics 249, Springer (1971), 75-83.

9. A. Sieradski, Square roots up to homotopy type, Amer. J. Math. 94 (1972), 73-81.

10. J. Stasheff, Manifolds of the homotopy type of (non-Lie) groups, Bull. A. M. S. 75 (1969), 998-1000.

11. D. Sullivan, Geometric topology, part I: Localization, periodicity and Galois symmetry, MIT, June 1970, (mimeographed notes).

12. A. Zabrodsky, Homotopy associativity and finite CW-complexes, Topology 9 (1970), 121-128.

MOD p DECOMPOSITIONS OF FINITE H - SPACES

by

John R. Harper*

Introduction

In this paper we study the mod p homotopy type of simply connected finite dimensional H - spaces. We shall use the notions of regularity [14] and quasi-regularity [10]. A prime p is said to be regular for a space X if there exists a product of spheres S and a map $f: S \longrightarrow X$ inducing mod p cohomology isomorphisms. Serre's result for compact Lie groups [14] has been extended to arbitrary finite H - spaces in the combined work of Browder [4] and Kumpel [6]. It reads; let X be a simply connected finite H - space with N denoting the largest entry in the type of X. If a prime p satisfies $2p - 1 \geq N$, then p is regular for X.

In a deeper analysis of the homotopy type of Lie groups, Mimura and Toda use the sphere bundles $B_n(p)$ and the idea of quasi-regularity. A prime p is said to be quasi-regular for a space K consisting of a

*Research supported in part by NSF grant GP-38024.

product of spheres and sphere bundles $B_n(p)$ and a map $f: K \to X$ inducing mod p cohomology isomorphisms. In [10], Mimura and Toda characterize the quasi-regular primes for the compact, simply connected, simple Lie groups. Naturally one wonders to what extent quasi-regularity depends only on the H-structure.

The main result of this paper is:

Theorem 1. Let X be a simply connected finite H-space with N the largest entry in its type. Assume X admits a multiplication making its rational cohomology primitively generated. If p is a prime satisfying $4p - 3 \geq N$, then p is quasi-regular for X.

This result agrees with the appropriate part of Theorem 4.2 of [10] except for the Lie groups F_4, E_i, $i = 6,8$. In these cases quasi-regularity holds for some primes $p \geq 5$ satisfying $4p - 3 < N$. Our method of proof shows that the reason is because of gaps in the types of these exceptional Lie groups. A general result can be formulated.

Theorem 2. Let X be as in Theorem 1 and assume further that $H_3(X;Z)$ is torsion free. Let p be a prime $p \geq 5$. A sufficient condition for quasi-regularity is $4p + 2q - 3 \geq N$ where q is some integer satisfying all of the following;

(a) $0 \leq q \leq p-1$,

(b) for all integers m such that $1 \leq m \leq q$ either

$2m+1$ or $4p+2m-3$ is missing from the type of X.

For example, in the case of F_4, the type is (3, 11, 15, 23). The $4p-3$ criterion yields $p = 7$. However taking $p = 5$ and $q = 3$ we get $4p+2q-3 = 23$. Condition (a) is satisfied. When $m = 2$ both 5 and 21 are missing. When $m = 3$, 7 is missing, so (b) is satisfied. Thus 5 is quasi-regular for F_4, which is the result in [10]. Together, Theorems 1 and 2 contain the sufficient conditions for quasi-regularity proved by Mimura and Toda.

By means of results of [11], the above Theorems can be formulated more geometrically. We denote by $X_{(p)}$ the localization of X at p. If p is quasi-regular for X (X as in Theorem 1 or 2) and $f:K \longrightarrow X$ is a mod p isomorphism then $f_{(p)}:K_{(p)} \longrightarrow X_{(p)}$ is a homotopy equivalence. This is immediate from [11] since the spaces involved are p-universal.

Our proof involves the Massey-Peterson spectral sequence [9]. While the arguments could be phrased purely in terms of Postnikov systems, this might obscure the structure of the argument. The crucial point is the observation that the mod p cohomology algebra of X

considered as an algebra over the Steenrod algebra completely determines the mod p homotopy type. This observation is most readily made via the spectral sequence -- or more precisely via the idea in the construction of the spectral sequence. In effect the machinery involved in setting up the spectral sequence reduces the proof to a routine, if somewhat tedious, algebraic calculation. The conditions of Theorem 2, while appearing rather involved, are readily seen to be those needed in order that the mod p homotopy type is determined by the cohomology structure.

Clarence Wilkerson has kindly informed me that he and Alex Zabrodsky have obtained results similar to Theorem 1. They also observe the relation between gaps in the type and anomalies to the $4p-3$ criterion. However their method of proof is quite different from the one given here.

Outline of the use of the Massey-Peterson spectral sequence.

In this section we outline the details of the construction and methods of computation of the Massey-Peterson spectral sequence. In [9] the spectral sequence is constructed for the prime 2. In broad outline the method parallels Adams [0] in establishing a connection between the homological algebra of resolutions and geometric realizations of resolutions as a first approximation to a Postnikov system. The theorem needed to establish the connection is the main result of [8] as extended in section 11 of [9]. For odd primes the analogous theorem has been proved by Barcus [1]. Thus for odd primes one can mimic part III of [9] to obtain the spectral sequence we use here. Different approaches to this spectral sequence appear in L. Smith [15] and Bousfield and Kan [2], [3].

We state the central fact on which our proofs are based. Let K denote a product of odd dimensional spheres S^{2m+1} and sphere bundles $B_n(p)$. Consider the case of Theorem 1. It will turn out that a K can be found with $2m+1 \leq 4p-3$ and $n \leq p-1$ such that $H^*(X; Z_p) \cong H^*(K; Z_p)$ abstractly as algebras over the Steenrod algebra. Thus the mod p cohomology of X has the form $\mathcal{U}(M)$ where M is an unstable A-module of a particularly nice form. Now the fact that emerges is that by inspection of E_2 of the spectral sequence there are no non-trivial differentials in stems $\leq 6p-5$. Furthermore all the extensions at E_∞ can be read off from E_2. This implies that if one writes down the geometric realization of a minimal resolution of M one obtains essentially the mod p Postnikov system of X through stages $6p-5$, as in Figure 1. The qualification is that a true geometric realization will not satisfy the connectivity conditions of a Postnikov system, but will be modified in the sense of [7]. In our situation, this point is easily handled using $K(Z, n)$ instead of $K(Z_p, n)$ at the appropriate place.

$$
\begin{array}{lll}
X & & \\
\downarrow & & \\
E_{6p-4} & & \\
\downarrow & & \\
E_{6p-5} & \longrightarrow & K(\Pi_{6p-4}, 6p-3) \\
\downarrow & & \\
E_{6p-6} & \longrightarrow & K(\Pi_{6p-5}, 6p-4) \\
\downarrow & & \\
\vdots & & \\
\downarrow & & \\
E_3 & \longrightarrow & K(\Pi_3, 4)
\end{array}
$$

Figure 1

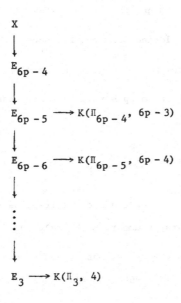

Since $n \leq p-1$, the dimension of $B_n(p)$ is $\leq 6p-4$. Moreover we will find that $H^*(E_{6p-5}, Z_p)$ and $H^*(X; Z_p)$ agree through dimensions $6p-4$. Hence a wedge of spheres and sphere bundles making up K can be mapped into E_{6p-5} and then lifted to X so that the induced map on mod p cohomology is epic. Since X is an H-space, the map can be extended to the product K so as to give the required cohomology isomorphism. The conditions in Theorem 2 permit the same argument, now in the range $10p-8$ although there is one bit of homotopy theory required. Thus the central fact is that in the needed range of dimensions the mod p Postnikov system of X is determined purely algebraically. While the spectral sequence is not absolutely necessary for such an argument, it organizes the computations and reveals the basic facts in a sharp straightforward manner.

REFERENCES

[0] J. F. Adams, On the structure and applications of the Steenrod algebra, Comment. Math. Helv. 32 (1958), 180-214.

[1] W. D. Barcus, On a theorem of Massey and Peterson, Quart. J. Math. 19 (1968), 33-41.

[2] A. K. Bousfield and D. M. Kan, The homotopy spectral sequence of a space with coefficients in a ring, Topology 11 (1972), 79-106.

[3] ______________ and ________, A second quadrant homotopy spectral sequence, Trans, Amer. Math. Soc. 176 (1973).

[4] W. Browder, Torsion in H - spaces, Ann. of Math. 74 (1961), 24-51.

[5] _________, On differential Hopf algebras, Trans. Amer. Math. Soc. 10 (1963), 153-176.

[6] P. G. Kumpel, Mod p equivalences of mod p H - spaces, Quart. J. Math. 23 (1972).

[7] M. Mahowald, On obstruction theory in orientable fibre bundles, Trans. Amer. Math. Soc, 110 (1964), 315-349.

[8] W. S. Massey and F. P. Peterson, The cohomology structure of certain fibre spaces, Topology 4 (1965), 47-65.

[9] ____________ and ______________, On the mod 2 cohomology Structure of certain fibre spaces, Amer. Math. Soc. Memoirs 74 (1967).

[10] M. Mimura and H. Toda, Cohomology operations and the homotopy of compact Lie groups, Topology 9 (1970), 317-336.

[11] _________ and _______, On p - equivalences and p - universal spaces, Comment. Math.

[12] S. Oka, The homotopy groups of sphere bundles over spheres, J. Sci. Hiroshima U. 33 (1969), 161-195.

[13] H. Samelson, Beiträge zur Topologie der Gruppen - Mannig - faltigkeiten, Ann. of Math. 42 (1941), 1091-1137.

[14] J. P. Serre, Groupes d'homotopie et classes des groups abeliens, Ann. of Math. 58 (1953), 258-294.

[15] L. Smith, Hopf fibration towers and the unstable Adams spectral sequence, Applications of Categorical Algebra, Proc. Sym. Pure Math. Providence, R. I. (1970).

[16] N. E. Steenrod and D. B. A. Epstein, Cohomology Operations, Ann. of Math. Studies 50 (1962), Princeton.

[17] H. Toda, Composition methods in homotopy groups of spheres, Ann. of Math. Studies 49 (1962), Princeton.

MOD p DECOMPOSITIONS OF MOD p H-SPACES

Clarence Wilkerson

(This was an expository talk explaining joint work with Alex Zabrodsky. John Harper has independently and simultaneously obtained similar results by somewhat different methods. Since complete proofs will appear elsewhere, the following is only a summary of the results.)

Definition: A finite CW complex X is a finite mod p H-space if its p-localization X_p is an H-space.

Theorem: (Hopf) If X is a finite mod p H-space, then

$$H^*(X,Q) = \Lambda(x_{2r_1-1},\cdots,x_{2r_n-1}).$$

We refer to the set $\{2r_1-1,\cdots,2r_n-1\}$ as the type of X and n as its rank. It is not hard to see that the theorem of Hopf implies that there is a map $F:S^{2r_1-1}\times\cdots\times S^{2r_n-1}\longrightarrow X$ such that $H^*(F,Q)$ is an isomorphism. If we make the convention that $1 < r_i \leq r_{i+1}$, then [Serre] and [Kumpel] have shown the following:

Theorem: If X is a finite mod p H-space with

$H^*(X,Z/p) = \Lambda(x_{2r_1-1},\cdots,x_{2r_n-1})$ and $r_n - r_1 < p-1$, then there exists a map $F:S^{2r_1-1}\times\cdots\times\cdots\longrightarrow X$ such that $H^*(F,Z/p)$ is an isomorphism.

The primes p for which there is a map F as above are called the regular primes of X. One of the simplest cases of a non-regular prime is the prime 3 for the Lie group $Sp(2)$. $Sp(2)$ has type $\{3,7\}$, but is not 3-equivalent to the product of S^3 with S^7. This shows that any extension of the Serre result must involve new factors other than spheres. In fact, the factors needed turn out to be analogues of $Sp(2)$.

Definition: $B_n(p)$ is the S^{2n+1} fibration over $S^{2n+1+2(p-1)}$ with characteristic map α_p, where α_p is the p-primary generator in $\pi_{2n+2(p-1)} S^{2n+1}$. $H^*(B_n(p), Z/p) = \Lambda(x_{2n+1}, P_p^1 x)$. For example, $Sp(2) =_3 B_1(3)$.

Using these additional factors, [Oka] and [Mimura-Toda] obtained further mod p decompositions of the Lie groups:

Theorem: If G is a simply connected Lie group such that $H_*(G,Z)$ has no p-torsion. Then if $r_n - 2 < 2(p-1)$, G is p-equivalent to a product of spheres and $B_{n_i}(p)$. The factors are determined by the action of P_p^1 on $H^*(G,Z/p)$.

The primes with the above property for G are called the quasi-regular primes of G. Theorem A below shows that the same result is valid for mod p H - spaces.

Theorem A: Let X be a simply connected mod p H - space with $H^*(X,Z/p) = \Lambda\left(x_{2r_1-1},\cdots,x_{2r_n-1}\right)$. If $r_n - r_1 < 2(p-1)$ and the $\left\{x_{2r_i-1}\right\}$ can be chosen such that either $P^1 x_{2r_i-1} = 0$ or x_{2r_j-1}, then X is p - equivalent to a product of spheres and $B_n(p)$'s.

We remark that the condition that the generators give a $Z/p(p^1)$ basis of the indecomposables is clearly true if the decomposition is possible, and hence Theorem A provides necessary and sufficient conditions for quasi-regularity in the range of dimensions $r_n - r_1 < 2(p-1)$. There are several possible approaches to the proof of Theorem A. The author's first approach was to prove that in this range of dimensions, any indecomposable Z/p - cohomology class not in the ideal generated by P^1 and the Bocksteins was necessarily spherical. This involved studying the connective tower of X, [Smith]. The $B_n(p)$ factors were mapped by a trick, the lifting theorem of [Zabrodsky], which allows one to conclude that a map defined through the dimensions of the indecomposables, automatically extends to a map of

$B_n(p)$. In higher ranges of dimensions, this approach becomes cumbersome.

The second approach is more conceptual. For each possible <u>Z/p - cohomology</u> operation in a p - torsion free H - space, we build a model, $X(\phi)$ as a two-stage (n - stage) Postnikov system. If the <u>Z/p - cohomology</u> of the mod p H - space X has the appropriate structure over the set of operations $\{\phi_i\}$, there is a map $X \longrightarrow \Pi_i X(\phi_i)$ which is a mod p equivalence through the dimensions of the indecomposables. Hence by the lifting theorem, X is p - equivalent to the product of the $\overline{X(\phi_i)}$, the finite dimensional approximations to $X(\phi_i)$ with cohomology an exterior algebra. In low ranges of dimensions in terms of $r_n - r_1$, the possible cohomology operations are easy to analyze. In particular, for $r_n - r_1 < 2(p-1)$, the only operation possible is P^1. This yields Theorem A. In the range $r_n - r_1 < 3(p-1)$, there are relatively few operations, and in addition to the $B_n(p)$, we obtain models with three generators in cohomology. Of course, the precise result becomes notationally difficult to state, but in particular, it is strong enough to cover the few cases of exceptional Lie groups which do not satisfy the hypothesis of Theorem A, but yet are quasi-regular anyway, see [Harper] also. Thus, given the cohomology operations, the [Mimura-Toda] results can be derived independently of any Lie group hypothesis.

References

Harper, J.: Quasi-regular primes for H-spaces, Preprint.

Kumpel, P.G.,Jr.: On p-equivalences of mod p H-spaces, Ouart. J. Math.,Oxford (II) 23 (1972), 173-178.

Mimura, M. and Toda, H.: Cohomology operations and homotopy of compact Lie groups, Topology 9 (1970), 317-336.

Oka, S.: On the homotopy groups of sphere bundles over spheres, J. Sci. Hiroshima U. 33 (1969(161-195.

Serre, J.P.: Groupes d'homotopie et classes de groupes abéliens, Ann. of Math. 58 (1953), 258-294.

Smith, L.: On the relation between spherical and primitive homology classes in topological groups, Topology 8 (1969), 69-80.

Wilkerson, C. and Zabrodsky, A.: Quasi-regular primes for mod p H-spaces, Preprint.

Zabrodsky, A: On rank two mod p odd H - spaces, Preprint.

ETALE HOMOTOPY THEORY AND SHAPE

by

David A. Edwards

I. Introduction

This article is a survey of some recent developments in the study of the algebraic topology of pathological spaces. The usual techniques and theorems of algebraic topology work well only when applied to spaces having the homotopy type of a CW - complex. For more pathological spaces, the singular theories completely break down, while the Cech theories retain much of their usefulness. As an example, consider the Warsaw circle S_w.

Fig. I.1.

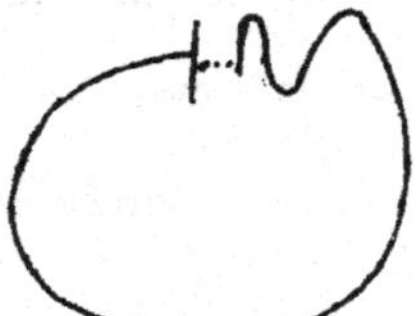

Globally S_w looks like the standard circle S ; but S_w has 'local' pathology. The singular homotopy groups of S_w all vanish, in particular $\pi_1(S_w) = 0$. On the other hand, the Cech fundamental group of S_w is equal to Z , i.e. $\check{\pi}_1(S_w) = \pi_1(S)$. In fact, $\check{F}(S_w) = F(S)$ for every functor F from the homotopy category of CW complexes $\mathcal{H}$ to the category of groups, where $\check{F}(S_w) = \varprojlim F(N)$ and the inverse limit is taken over nerves of coverings of S_w. This is easily seen by oberving that the Cech tower of nerves of coverings of S_w has a co-final sub-tower each element of which is a simplicial complex having the homotopy type of a circle and the bonding maps all have degree one.

Unfortunately, Cech homology is known not to be exact in general. This lack of exactness is due to the fact that the inverse limit functor is not exact. One way of circumventing this problem is not to take inverse limits but to learn to work with towers. The Cech construction associates a tower of CW complexes to a space. What one needs is to form a category of towers such that co-final towers are isomorphic. Such a construction is originally due to Grothendieck [19] and will be described in Section II. Grothendieck's construction can be applied to any category C to yield a category pro - C. The Cech construction defines the Cech functor C:Top ⟶ Pro - H and an analogous construction defines the etale functor E:Schemes ⟶ Pro - H. The algebraic topology of Pro - H will be described in Section II. In Section III we survey Etale Homotopy Theory and in Section IV we survey Cech Homotopy Theory (better known as Shape Theory). Section V is concerned with classification.

II. Pro - H.

The need for Pro- H has now arisen in four different fields.

1. Algebraic Geometry: In algebraic geometry it serves as the

range category for the étale functor $\mathcal{E}$:Schemes $\longrightarrow$ Pro - $\mathcal{H}$. $\mathcal{E}$ was originally defined by Grothendieck in an attempt to carry out Weil's program for proving the Weil conjectures. This program has recently been successfully completed by Deligne. Artin and Mazur [3] also use $\mathcal{E}$ to compare schemes defined over different rings. They prove a comparison theorem which was used by Quillen [35] in his 'proof' of the Adams' conjecture.

2. Algebraic Topology: In algebraic topology one can study the homotopy type of a CW - complex X by 'fracturing' it into rational and mod P components $(X_0, X_2, X_3, \cdots)$. The homotopy type of X can then be recovered from $(X_0, X_2, X_3, \cdots)$ together with coherence information over the rationals. This point of view is due mainly to Serre (mod C Theory [36]) and Sullivan [38]. The recent book by Bousfield and Kan [8] presents a thorough treatment of these ideas. The idea of the pro-finite type of a CW - complex first appeared in Artin and Mazur [3] where it is a pro-object. Sullivans' pro-finite type is an inverse limit of the Artin-Mazur pro-finite type. Pro - $\mathcal{H}$ is also the natural setting for the study of Postnikoff systems.

3. Geometric Topology: In geometric topology Sullivan and others use the formalism of completions and localizations to describe the homotopy type of the spaces G/PL, G/TOP, etc.

Sullivan's proof of the Adams' conjecture and its generalizations also uses the above formalism.

4. General Topology: In general topology Borsuk [5] studies closed subsets of the Hilbert cube Q by associating to each such subset of Q its fundamental sequence of open neighborhoods. Thus, to each $X \subset Q$ Borsuk associates its shape $SH(X) \in \text{Pro}-\mathcal{H}ANR$, where $\mathcal{H}ANR$ is the homotopy category of ANRS (= absolute neighborhood retracts). Borsuk and his school then go on to develop the algebraic topology of $\text{Pro}-\mathcal{H}ANR$ and to classify continua up to shape (i.e., $X \underset{SH}{\simeq} Y$ iff $SH(X) \cong SH(Y)$ in $\text{Pro}-\mathcal{H}ANR$). More generally, one can use the Cech construction to define a functor $C: TOP \longrightarrow \text{Pro}-\mathcal{H}$.

One should note the similarity between the algebraic geometer's use of the étale functor E and the general topologist's use of the Cech functor C. In both cases one has a Cech - like construction to associate to a pathological object a tower of CW - complexes,

II.1

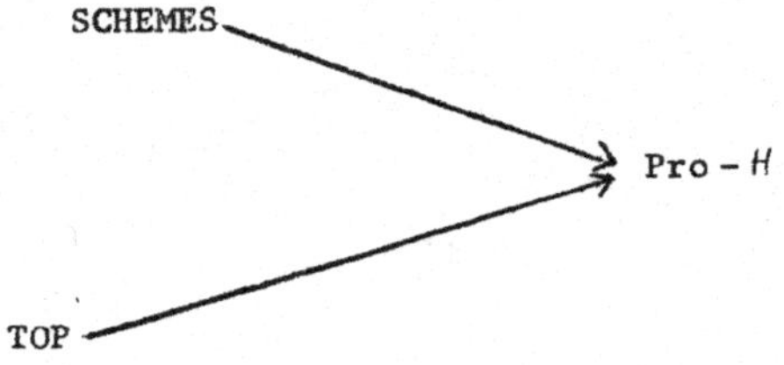

Each theory then breaks up into two parts. First, one must extend the usual results of algebraic topology from $\mathcal{H}$ to Pro - $\mathcal{H}$. This has been carried out by Artin and Mazur [3]. The general topologists have proved some interesting results about Pro - $\mathcal{H}$ not contained in [3]. For example, Moszynska [32] has proved a Pro - Whitehead Theorem for towers of CW - complexes of finite homotopy dimension. The second part of each theory concerns the classification of objects in SCHEMES or in TOP up to isomorphism of their images in Pro - $\mathcal{H}$. Here we have the comparison theorems of Artin and Mazur [3] and the shape classification theorems of Chapman [9], Edwards and Geoghegan [11], Keesling [23], Mardesic [28] and others.

Now for the formal definition of Pro - $\mathcal{H}$. In order to perform certain constructions such as pro-finite completion one needs a very general definition of an inverse system.

Def. II.1: A category I is said to be filtering if:

a) (directedness) Every pair i,i' of objects of I can be embedded in a diagram

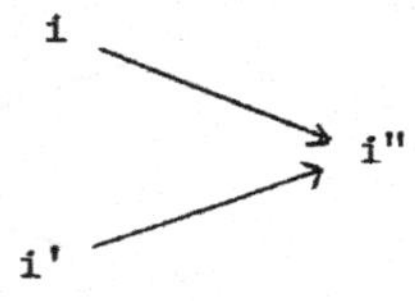

;

b) (essential uniqueness) If $i \rightrightarrows i'$ is a pair of maps of I, there is a map $i' \longrightarrow i''$ such that the two compositions are equal.

Def. II.2: Let C be a category. An inverse system in C is a contravariant functor $X: I^{\circ} \longrightarrow C$ whose domain category I is filtering.

Def. II.3: Let C be a category. $\text{Pro}-C$ is the category whose objects are inverse systems $X = \{X_i\}_{i \in I}$ in C and whose set of morphisms from $X = \{X_i\}_I$ to $Y = \{Y_\gamma\}_J$ is

$$\text{Pro}-C\,(X,Y) \equiv \underset{j \in J}{\underleftarrow{\text{Lim}}}\ \underset{i \in I}{\underrightarrow{\text{Lim}}}\ C(X_i, Y_j)\ .$$

Note that the indexing categories are not assumed equal. We have defined the set of maps in $\text{Pro}-C$ from X to Y, but the above definition is somewhat opaque and it's not obvious how to define the composition of two maps from the above definition. Hence, we shall give an alternative definition. For simplicity, assume that we are given inverse systems $\{X_i\}_I$ and $\{Y_\gamma\}_J$ in C which are indexed by directed sets I and J.

Def. II.4: A morphism $f: X \longrightarrow Y$ in $\text{Pro}-C$ is represented by a map $\theta: J \longrightarrow I$ (not necessarily order preserving) and morphisms $f_j: X_{\theta(j)} \longrightarrow Y_j$ of C for each $j \in J$, subject to the condition that if $j \leq j'$ in J then for some $i \in I$ such that $i \geq \theta(j)$ and $i \geq \theta(j')$, the diagram

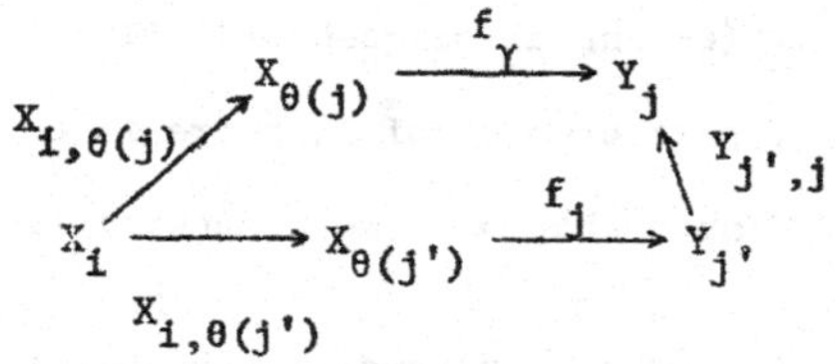

commutes ($x_{i,i'}:X_i \longrightarrow X_i$ and $Y_{j,j'}:Y_h \longrightarrow Y_{j'}$ are the structure maps of the inverse system). Two pairs (θ, f_j) and (θ', f'_j) represent the same morphism in $\text{Pro}-\mathcal{C}$ if for each $j \in J$ there is an $i \in I$ such that $i \geq \theta(j)$ and $i \geq \theta'(j)$ and $f_j X_{i,\theta(j)} = f'_j X_{i,\theta'(j)}$.

Remark: See Fox [14] for a very lucid description of morphisms in $\text{Pro}-\mathcal{H}$. The pro-object $\{X_i\} \in \text{Pro}-\mathcal{C}$ contains much more information about the inverse system than does the inverse limit $\varprojlim X_i \in \mathcal{C}$ even if the inverse limit exists in $\mathcal{C}$ - it might not exist in $\mathcal{C}$. The relationship between the pro-object $\{X_i\}$ and the inverse limit $\varprojlim X_i$ is analogous to that between the germ of a function f at a point p and the value of f at p. For more details concerning $\text{Pro}-\mathcal{C}$ see [3].

The basic homotopy categories we require are described in the following definitions.

Def. II.5: 1. $\mathcal{H}$ ≡ The homotopy category of CW-complexes;

2. $\mathcal{H}_0$ ≡ The pointed homotopy category of connected pointed CW-complexes;

3. $\mathcal{H}_{0,\text{pairs}}$ ≡ The pointed homotopy category of connected CW-pairs. More precisely, an object of $\mathcal{H}_{0,\text{pairs}}$ is an actual pointed map of pointed CW-complexes $X \xrightarrow{f} Y$ and a morphism of $\mathcal{H}_{0,\text{pairs}}$ is a homotopy equivalence class of actual commutative diagrams

$$\begin{array}{ccc} X' & \xrightarrow{f'} & Y' \\ \downarrow g_1 & & \downarrow g_2 \\ X & \xrightarrow{f} & Y \end{array} \quad ;$$

4. Pro-$\mathcal{H}$, Pro-$\mathcal{H}_0$, Pro-$\mathcal{H}_{0,\text{pairs}}$;

5. The analogous semi-simplicial categories:

$\mathcal{K}, \mathcal{K}_0, \mathcal{K}_{0,\text{pairs}}$, Pro-$\mathcal{K}$, Pro-$\mathcal{K}_0$, Pro-$\mathcal{K}_{0,\text{pairs}}$.

Note: The subscript 0 will be used to indicate pointed connected objects.

If $T: \mathcal{C} \longrightarrow \mathcal{A}$ is a functor, then T extends to a functor Pro-T: Pro-$\mathcal{C} \longrightarrow$ Pro-$\mathcal{A}$ defined by Pro-T $(\{X_i\}) = \{TX_i\}$. We thus obtain extensions of the usual functors of algebraic topology, in particular for homology and homotopy we have

<u>Def. II.6</u>: 1. $\text{Pro}-H_n : \text{Pro}-\mathcal{H} \longrightarrow \text{Pro}-AB \equiv$ Pro-abelian groups

$$\{X_i\} \longmapsto \{H_n(X_i)\};$$

2. $\text{Pro}-\pi_1 : \text{Pro}-\mathcal{H}_0 \longrightarrow$ Pro-groups

$$\{X_i\} \longmapsto \{\pi_1(X_i)\};$$

3. $\text{Pro}-\pi_n : \text{Pro}-\mathcal{H}_0 \longrightarrow \text{Pro}-AB, \quad n > 1$

$$\{X_i\} \longmapsto \{\pi_n(X_i)\}.$$

Since $\underrightarrow{\text{Lim}}$ is an exact functor, it doesn't hurt to take the limit when extending contravariant functors from $\mathcal{H}$ to $\text{Pro}-\mathcal{H}$. In particular, for cohomology we have

<u>Def. II.7</u>: $H^n : \text{Pro}-\mathcal{H} \longrightarrow AB$

$$\{X_i\} \longmapsto \underrightarrow{\text{Lim}}\, H^n(X_i).$$

Let X be a CW-complex and $\text{cosk}_n X$ be the CW-complex obtained from X by killing all the homotopy groups in dimensions greater than n. This can be done functorially using semi-simplicial theory (see [3]). Thus, we have functors

$$\text{cosk}_n : \mathcal{H}_0 \longrightarrow \mathcal{H}_0$$

and their extensions

$$\text{cosk}_n : \text{Pro}-\mathcal{H}_0 \longrightarrow \text{Pro}-\mathcal{H}_0 .$$

These functors combine to yield a functor $\natural$.

Def. II.7: $\natural : \text{Pro}-\mathcal{H}_0 \longrightarrow \text{Pro}-\mathcal{H}_0$

$$X = \{X_i\}_i \longmapsto X^\natural = \{\text{cosk}_n\, X_i\}_{n,i} .$$

$X^\natural$ is a canonical Postnikoff decomposition for X.

Def. II.8: A map $f:X \longrightarrow Y$ in $\text{Pro}-\mathcal{H}_0$ is said to be a $\natural$-isomorphism if $f^\natural : X^\natural \longrightarrow Y^\natural$ is an isomorphism.

Theorem II.1 ([3]): f is a $\natural$-isomorphism iff $\text{Pro}-\pi_n(f) : \text{Pro}-\pi_n(X) \longrightarrow \text{Pro}-\pi_n(Y)$ is an isomorphism in pro-groups for all n .

Remark: Thus a $\natural$-isomorphism can also be called a weak homotopy equivalence. The Whitehead Theorem is not true in general in $\text{Pro}-\mathcal{H}_0$, i.e. $\text{Pro}-\pi_n(f)$ may be an isomorphism for all n but f may fail to be an isomorphism. For example, the natural map $S^2 \longrightarrow S^{2\natural}$ is a $\natural$-isomorphism but not an isomorphism in $\text{Pro}-\mathcal{H}_0$. A more striking example is the following. Let $S_n^\infty = \bigvee_{k>n} S^k$ and $S^\infty = \{S_n^\infty\}$ where the bonding maps are the obvious inclusions. S^∞ is $\natural$-isomorphic to a point but not isomorphic to a point. (See [3] and [11]).

Let $\pi_*(X) \equiv \prod_{k \geq 1} \pi_k(X)$.

Conjecture II.1: (Whitehead Theorem in $\text{Pro}-\mathcal{H}_0$) $f:X \longrightarrow Y$ in $\text{Pro}-\mathcal{H}_0$ is an isomorphism if and only if $\text{Pro}-\pi_*(f)$ is an isomorphism in Pro-groups.

We shall now define the notion of the homotopy dimension of an object in $\text{Pro}-\mathcal{H}$.

Def. II.9: 1. $\text{Dim}\,\{X_i\} = \text{Sup}\,\{\text{Dim}\,X_i\}$.

2. $H-\text{Dim}\,\{X_i\} = \text{Inf}\,\{\text{Dim}\,\{Y_i\} \mid \{Y_i\} \underset{\text{Pro}-\mathcal{H}_0}{\simeq} \{X_i\}\}$.

Let $\mathcal{H}_{0,f}$ be the subcategory of $\mathcal{H}_0$ consisting of finite CW-complexes.

Theorem II.2 (Moszynska [32]): (Whitehead Theorem in $\text{Pro}-\mathcal{H}_{0,f}$) Let $X,Y \in \text{Pro}-\mathcal{H}_{0,f}$ have finite homotopy dimension. Then $f:X \longrightarrow Y$ is an isomorphism iff $\text{Pro}-\pi_n(f)$ is an isomorphism for all n.

Theorem II.3 ([3]): (Hurewicz Theorem in $\text{Pro}-\mathcal{H}_0$) Let $X \in \text{Pro}-\mathcal{H}_0$ and suppose that $\text{Pro}-\pi_q(X) = 0$ for $q < n$ where $n > 1$. Then the canonical map

$$h : \text{Pro}-\pi_n(X) \longrightarrow \text{Pro}-H_n(X)$$

is an isomorphism.

The analogues in $\text{Pro}-\mathcal{H}_0$ of many maps which are isomorphisms in $\mathcal{H}_0$ turn out to be only $\natural$-isomorphisms. For example,

Theorem II.4)[3]): (Uniqueness of Eilenberg-MacLane Pro-spaces) Let $X \in \text{Pro}-\mathcal{H}_0$ and suppose $\text{Pro}-\pi_q(X) = 0$ for $q \neq n$ and $\text{Pro}-\pi_n(X) = G = \{G_i\} \in \text{Pro}-\text{groups}$. Then X is $\natural$-isomorphic to $K(G,n) = \{K(G_i,n)\}$.

We have now described the main results concerning the algebraic topology of $\text{Pro}-\mathcal{H}_0$. For more details see [3].

III. Etale Homotopy Theory

The Cech construction defines a functor $C:\text{Top} \longrightarrow \text{Pro}-\mathcal{H}$. If $X \in \text{Top}$, then $C(X)$ is called the Cech homotopy type of X. If X is a connected complex algebraic variety topologized by the Zariski topology, then $C(X) \simeq$ point in $\text{Pro}-\mathcal{H}$. This is due to the fact that every Zariski open set is dense, and hence the nerve of any finite covering of X is a simplex. So, if one wants to study algebraic varieties and schemes one must be more subtle. One is led to generalize the notion of a topology for a space. If $X \in \text{Top}$, define $T(X)$ to be the category whose objects are open embeddings $\phi:U \longrightarrow X$ and whose morphisms are commutative diagrams

$$\begin{array}{ccc} U & \xrightarrow{\chi} & U' \\ & {\scriptstyle\phi}\searrow \quad \swarrow{\scriptstyle\phi'} & \\ & X & \end{array}$$

A collection $\{\phi_\alpha:U_\alpha \longrightarrow U\}$ of morphisms of $T(X)$ is a covering of U if the images of the ϕ_α cover U. $T(X)$ is the ordinary Grothendieck topology associated to a topological space X. We are thus led to make the following definition.

Def. III.1: A Grothendieck topology τ on a category C consists of a category $C = \text{Cat}\,\tau$ and a set $\text{Cov}\,\tau$ of families $\{U_i \xrightarrow{\phi_i} U\}_{i \in I}$ of maps in $\text{Cat}\,\tau$ called coverings (where in each covering the range U of the maps ϕ_i is fixed) satisfying

1. If ϕ is an isomorphism, then $\{\phi\} \in \text{Cov}\,\tau$.
2. If $\{U_i \longrightarrow U\} \in \text{Cov}\,\tau$ and $\{V_{ij} \longrightarrow U_i\} \in \text{Cov}\,\tau$ for each i, then the family $\{V_{ij} \longrightarrow U\}$ obtained by composition is in $\text{Cov}\,\tau$.
3. If $\{U_i \longrightarrow U\} \in \text{Cov}\,\tau$ and $V \longrightarrow U \in \text{Cat}\,\tau$ is arbitrary, then $U_i \underset{U}{\times} V$ exists and $\{U_i \underset{U}{\times} V \longrightarrow V\} \in \text{Cov}\,\tau$.

If τ is a Grothendieck topology with an initial object (to play the role of the empty set) and a terminal object (to play the role of the total space), then one can apply the Cech construction to τ and obtain a Pro-object in Pro-$\mathcal{H}$. One thus obtains an extension of the Cech functor to the category $\mathcal{G}$ of Grothendieck topologies containing initial and terminal objects, and a commutative diagram

III.1
$$\begin{array}{ccc} G & \xrightarrow{\ C\ } & \text{Pro} - H \\ & {\scriptstyle i}\nwarrow \quad \nearrow{\scriptstyle C} & \\ & \text{Top} & \end{array} ,$$

where i sends $X \in \text{Top}$ to $\mathcal{T}(X) \in G$. This still ends up being inadequate for the purposes of algebraic geometry. If $\{U_\alpha \xrightarrow{\phi_\alpha} U\} \in \text{Cov}\ \tau$, then one forms the following simplicial object of Cat τ. Let $V = \coprod U_\alpha \xrightarrow{\phi = \coprod \phi_\alpha} U$, then consider the simplicial object.

III.2
$$V \leftleftarrows V \underset{U}{\times} V \leftleftarrows V \underset{U}{\times} V \underset{U}{\times} V \leftleftarrows \cdots$$

III.2 is the basic construction of Cech Theory.

Def. III.2: A hypercovering $X.$ of a Grothendieck Topology $\tau \in G$ is a simplicial object with values in τ satisfying the following conditions for all n:

($SURJ_0$). Let e be the final object of τ. The map $X_0 \longrightarrow e$ is a covering.

($SURJ_n$). The canonical morphism $X_{n+1} \longrightarrow (\text{Cosk}_n X.)_{n+1}$ is a covering.

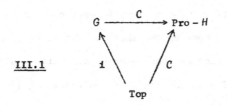

III.2 is a hypercovering. The notion of hypercovering allows us to pass to refinements as we move up the skeleta. 'It is this option of refining further in each dimension that makes hypercoverings useful in cases where the category of mere coverings is too coarse'. [3]

<u>Theorem III.1</u> ([3]): Let $\tau \in G$ and $HR(\tau)$ denote the category whose objects are hypercoverings, and whose maps are homotopy classes of morphisms. Then the opposite category $HR(\tau)^{o}$ is filtering.

Thus, any functor $\tau \xrightarrow{\pi}$ sets will induce a functor $HR(\tau) \xrightarrow{\pi_*} K =$ homotopy category of simplicial sets, by sending the hypercovering $X_.$ to the simplicial set $\pi \circ X.$. Since $HR(\tau)^{o}$ is filtering, π_* is an object in Pro - K. Using the geometric realization functor we can pass to Pro - H. The functor π considered by Artin and Mazur assigns to every $A \in \tau$ its set of connected components, where τ is assumed to be locally connected (see [3]). (It is here that the Verdier construction fails to be useful in general topology since most pathological spaces of interest are definitely not locally connected.) Let $G_{\ell c}$ be the subcategory of G consisting of locally connected topologies. Then, we have the Verdier functor

III.3: $\prod : G_{\ell c} \longrightarrow \text{Pro} - H.$

For $\tau \in G_{\ell c}$, $\pi(\tau)$ and $C(\tau)$ are not in general isomorphic. In fact, there is a connected, locally connected four point topological space X such that $\pi(X) \neq C(X)$.

Problem III.1: Let $\mathrm{Metric}_{\ell c} \overset{\pi}{\underset{C}{\rightrightarrows}} \mathrm{Pro}-H$ be the restrictions of π and C to $\mathrm{Metric}_{\ell c}$. Are these functors naturally equivalent.

If X is a scheme, then there are several natural Grothendieck topologies one can associate with X. In particular, besides the Zariski topology $Z(X)$, one also has the étale topology $E(X)$. One takes as objects of the étale topology on X not only the Zariski open subsets U of X but also surjective étale mappings $V \longrightarrow U$. An étale map $V \xrightarrow{f} U$ should be thought of as a finite covering space - over the complex numbers every étale map determines a finite covering space. (See [24] for more details concerning Grothendieck topologies and the étale topology.) Thus, coverings in the étale topology are both horizontal and vertical.

Def. III.3: Let $X \in \mathrm{Schemes}_{\ell c}$ = category of locally connected schemes. Then define the etale homotopy type of X, $X_{\ell c} \in \mathrm{Pro}-H$, as the Pro-object

$$\mathrm{HR}(E(X)) \xrightarrow{\pi_*} K \xrightarrow{|\cdot|} H .$$

We thus obtain a functor

$$E:\mathrm{Schemes}_{\ell c} \longrightarrow \mathrm{Pro}-H .$$

Def. III.4: A (Serre) class C of groups is a full subcategory of the category of groups satisfying:

1. $0 \in C$, i.e., the trivial group 0 is in C ;
2. Any subgroup of a C-group is a C-group. Moreover, if $0 \longrightarrow A \longrightarrow B \longrightarrow C \longrightarrow 0$ is an exact sequence of groups, then $B \in C$ iff $A,C \in C$.

C is called complete if in addition to 1. and 2. We have

3. If $A,B \in C$, then the product A^B of A with itself indexed by B is in C.

Examples: 1. The class of finite groups;

2. The class of finite groups whose orders are products of primes coming from a given set ℓ of primes.

Theorem III.2 ([3]): Let C be a class of groups. The inclusion of Pro $-$ C into Pro $-$ groups has an adjoint

$$\wedge : \text{Pro} - \text{groups} \longrightarrow \text{Pro} - C .$$

If $G \in$ Pro $-$ groups, then $\hat{G}$ is called the C-completion of G.

One way of describing $\hat{G}$ is as follows.

<u>Def. III.5</u>: Let $(G \downarrow \mathcal{C})$ be the category of $\mathcal{C}$-groups under G, i.e., an object of $(G \downarrow \mathcal{C})$ is a map $G \xrightarrow{\phi} C$, where $C \in \mathcal{C}$, and a morphism of $(G \downarrow \mathcal{C})$ is a commutative diagram

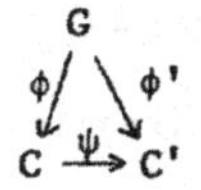

<u>Theorem III.3 ([3])</u>: $(G \downarrow \mathcal{C})^o$ is filtering.

<u>Proof</u>: Condition 2 of Def. III.4 is the essential ingredient. If $\phi: G \longrightarrow C$ and $\phi': G \longrightarrow C'$ with $C, C' \in \mathcal{C}$, then by Def. III.4-2, $C \pi C' \in \mathcal{C}$ and we have

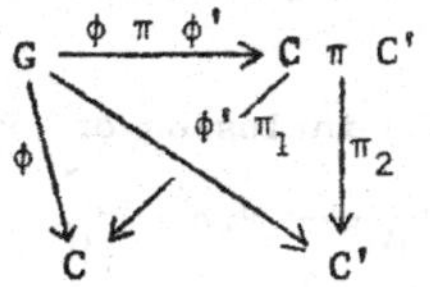

Hence $(G \downarrow \mathcal{C})^o$ is directed. If

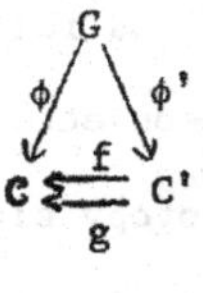

is commutative with $C, C' \in \mathcal{C}$, then the equalizer $E(f,g)$ of f and g is in $\mathcal{C}$ by Def. III.4-2 and we have

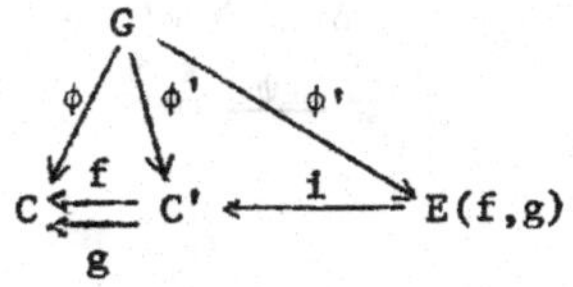

Hence $(G \downarrow \mathcal{C})^{\circ}$ has essential uniqueness, and is thus filtering.

<u>Def. III.6</u>: Define the forgetful functor $(G \downarrow \mathcal{C}) \longrightarrow \mathcal{C}$ by sending $(G \xrightarrow{\phi} C) \in (G \downarrow \mathcal{C})$ to $C \in \mathcal{C}$. $(G \downarrow \mathcal{C}) \longrightarrow \mathcal{C}$ defines an object $\hat{G}$ in $\text{Pro}-\mathcal{C}$. We thus obtain a functor $\wedge : \text{Pro}-\text{groups} \longrightarrow \text{Pro}-\mathcal{C}$ which is adjoint to the inclusion $\text{Pro}-\mathcal{C} \longrightarrow \text{Pro}-\text{groups}$.

The above construction and its variants are central to this field of mathematics.

<u>Def. III.7</u>: Let $\mathcal{C}$ be a class of groups. Define $\mathcal{CH}_0$ to be the full subcategory of $\mathcal{H}_0$ consisting of pointed connected CW-complexes whose homotopy groups are all in $\mathcal{C}$.

<u>Theorem III.4 ([3])</u>: The inclusion of $\text{Pro}-\mathcal{CH}_0$ into $\text{Pro}-\mathcal{H}_0$ has an adjoint $\wedge : \text{Pro}-\mathcal{H}_0 \longrightarrow \text{Pro}-\mathcal{CH}_0$. If $X \in \text{Pro}-\mathcal{H}_0$, then $\hat{X} \in \text{Pro}-\mathcal{CH}_0$ is called the $\mathcal{C}$-completion of X.

We can describe $\hat{X}$ as follows. Let $X \in \mathcal{H}_0$ and let $(X \downarrow \mathcal{CH}_0)$ be the category of $\mathcal{CH}_0$-objects under X, i.e. an object of $(X \downarrow \mathcal{CH}_0)$ is a map (homotopy class of) $X \xrightarrow{\phi} W$, where $W \in \mathcal{CH}_0$, and a morphism $\psi : \phi \longrightarrow \phi'$ is a commutative triangle

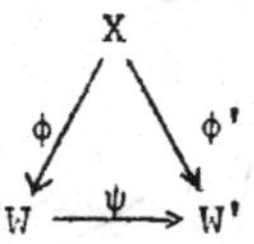

<u>Theorem III.5 ([3])</u>: $(X \downarrow \mathcal{CH}_0)^o$ is filtering.

Hence, the forgetful functor

$$(X \downarrow \mathcal{CH}_0) \longrightarrow \mathcal{CH}_0$$

$$(X \xrightarrow{\phi} W) \longmapsto W$$

defines an object X in $\text{Pro}-\mathcal{CH}_0$. We thus obtain completion functors

$$\wedge: \mathcal{H}_0 \longrightarrow \text{Pro}-\mathcal{CH}_0$$

$$\wedge: \text{Pro}-\mathcal{H}_0 \longrightarrow \text{Pro}-(\text{Pro}-\mathcal{CH}_0) = \text{Pro}-\mathcal{CH}_0 ,$$

with $\wedge: \text{Pro}-\mathcal{H}_0 \longrightarrow \text{Pro}-\mathcal{CH}_0$ adjoint to the inclusion $\text{Pro}-\mathcal{CH}_0 \longrightarrow \text{Pro}-\mathcal{H}_0$.

<u>Remark</u>: Mardesic [29] has used a variant of the above to define a shape functor from all of **Top** to a shape category $\mathcal{S}$. If $X \in \text{Top}$, let $(X \downarrow \mathcal{H})$ be the homotopy category of CW-complexes under X. The objects of $\mathcal{S}$ are topological spaces and the morphisms of $\mathcal{S}$ from X to Y are functors $(Y \downarrow \mathcal{H}) \longrightarrow (X \downarrow \mathcal{H})$. $(X \downarrow \mathcal{H})$ is filtering ([39] or [25]) and hence determines an object $((X \downarrow \mathcal{H}) \longrightarrow \mathcal{H}) = \mathcal{M}(X)$ in $\text{Pro}-\mathcal{H}$. Let $C_n(X)$ be the Cech homotopy type of X based upon numerable coverings. There is a natural equivalence $C_n \longrightarrow \mathcal{M}$ ([39],[25]). Note that C and C_n are not equivalent off paracompact spaces, in particular there exists an $X \in \text{Top}$ such that $C_n(X) \simeq pt$ while $C(X) \simeq S^1$ in $\text{Pro}-\mathcal{H}$.

Let $A \in H_0$ and S^nA denote its nth iterated reduced suspension.

Def. III.8: The stable homotopy category SH_0 has as objects pointed connected CW-complexes, and as morphisms from A to B the abelian group

$$SH_0(A,B) \equiv \{A,B\} \equiv \varinjlim H_0(S^nA,S^nB).$$

Let $\pi_n^S(A) = \{S^n,A\}$ = the nth-stable homotopy group of A and for $A = \{A_i\} \in \text{Pro}-SH_0$, define $\text{Pro}-\pi_n^S(A) = \{\pi_n^S(A_i)\}$. Let $\wedge$ denote profinite completion, i.e., C-completion for C = class of finite groups.

Theorem III.6 ([3]): Let X be a finite CW-complex. Then the natural map $X \longrightarrow \hat{X}$ induces profinite completions $\pi_1(X) \longrightarrow \pi_1(\hat{X})$ and $\pi_n^S(X) \longrightarrow \pi_n^S(\hat{X})$.

The general situation is fairly complicated with respect to π_n, but one does have

Theorem III.7 ([3]): Let C be a class of finite groups and $r \geq 1$ an integer. A Pro-abelian group G is C good iff the map $\widehat{K(G,r)} \longrightarrow K(\hat{G},r)$ is a $\natural$-isomorphism. If $X \in \text{Pro}-H_0$ is simply connected, then $\widehat{\pi_q(X)} \xrightarrow{\simeq} \pi_q(\hat{X})$ for $q \leq n$ if $\pi_q(X)$ is C-good for $q \leq n-1$.

Def. III.9: $A,B \in H_0$ are said to be of the same stable Prohomotopy type if A is isomorphic to B in $\text{Pro}-SH_0$.

Theorem III.8 ([3]): Let X be a finite CW-complex. There is only a finite number of stably inequivalent CW-complexes of the same stable prohomotopy type as X.

Now for some theorems concerning the étale homotopy type of schemes.

Theorem III.9 ([3]): (Profiniteness of Schemes): Let X be a pointed, connected, geometrically unibranch (e.g. normal), noetherian prescheme. Then the étale homotopy type of X, X_{et}, is pro-finite, i.e., is isomorphic to its profinite completion $\hat{X}_{et} \in \text{Pro}-\mathcal{CH}_0$.

Theorem III.10 ([3]): (Generalized Riemann Existence Theorem) Let X be a connected, pointed prescheme of finite type over the field of complex numbers. Then there is a canonical map $\epsilon: X_{CL} \longrightarrow X_{et}$, and it induces an isomorphism on profinite completions.

Note: X_{CL} is the classical homotopy type of X as a complex algebraic variety.

Theorem III.11 ([3]): If X is also geometrically unibranch, then X_{et} is isomorphic to the profinite completion of X_{CL}.

Theorem III.12 ([3]): Let k be a field admitting two embeddings $\epsilon_1, \epsilon_2: k \longrightarrow \mathbb{C}$ into the complex numbers, and let X be a pointed scheme of finite type over k, X_i the schemes over $\mathbb{C}$ obtained by the embeddings ϵ_i, i = 1,2. Then $\hat{X}_{1,CL} \simeq \hat{X}_{2,CL}$.

<u>Theorem III.13 ([3])</u>: Let R be a discrete valuation ring with separably algebraically closed residue field k, and let $f:X \longrightarrow \mathrm{Spec}(R)$ be a smooth proper scheme with connected geometric fibres X_0, X_1, both being assumed pointed compatibly with a chosen section of $X/\mathrm{Spec}(R)$. Then there is a canonical isomorphism $\hat{X}_{0,et} \simeq \hat{X}_{1,et}$, where $\wedge$ denotes completion with respect to the class of finite groups of order prime to the characteristic p of k.

<u>Remark</u>: Comparison theorems of the above type have been successfully used by Quillen [35] (see also [15]) to settle the Adams' conjecture. In particular, Quillen uses the following

<u>Theorem III.14 ([35])</u>: Let R be a strict localization of Z(= the integers) at the prime p and choose an embedding $R \longrightarrow \mathbb{C}$ (= complex numbers). The residue field k of R is an algebraically closed field of characteristic p. Let $X \longrightarrow \hat{X}$ denote C-completion, where C is the class of finite groups of order prime to p. Let V_R be a prescheme over Spec R with a given rational point $\mathrm{Spec}(R) \longrightarrow V_R$, and let $V_{\mathbb{C}}$ (resp. V_k) be the geometrically pointed prescheme obtained from V_R by base extension relative to the map $\mathrm{Spec}(\mathbb{C}) \longrightarrow \mathrm{Spec}(R)$ (resp. $\mathrm{Spec}(k) \longrightarrow \mathrm{Spec}(R)$). If V_R is proper and smooth over Spec(R) and if it is simply connected, then there are

isomorphisms in $\text{Pro}-CH_0$

$$\hat{V}_{C,CL} \xrightarrow{\epsilon} \hat{V}_{C,et} \xrightarrow{j} \hat{V}_{R,et} \xleftarrow{i} \hat{V}_{k,et},$$

where i and j are base change morphisms and where ϵ comes from the canonical map from the classical to the étale topology.

IV. Shape

The Cech construction defines functors:

IV.1 1. $C:\text{Top} \longrightarrow \text{Pro}-H$;

2. $C_0: \text{Top}_0 \longrightarrow \text{Pro}-H_0$;

3. $C_{0,\text{Pairs}}:\text{Top}_{0,\text{Pairs}} \longrightarrow \text{Pro}-H_{0,\text{Pairs}}$.

$C_{0,\text{Pairs}}$ is defined as follows. Let $f:X \longrightarrow Y$ be a pointed map in Top_0. Let $\{U_\alpha\}_A$ be a pointed open covering of Y and $\{U_\beta\}_B$ a pointed open covering of X and $\nu:B \longrightarrow A$ a pointed map such that $V_\beta \subset f^{-1}(U_{\nu(\beta)})$. ν determines a simplicial map $N(\nu)$ from the nerve $N\{V_\beta\}$ to the nerve $N\{U_\alpha\}$, $N(\nu):N\{V_\beta\} \longrightarrow N\{U_\alpha\}$. Letting $\{U_\alpha\}$ vary over all coverings of Y and $(\{V_\beta\},\nu)$ over all refinements of $\{f^{-1}(U_\alpha)\}$, we obtain the object

$$C_{0,\text{Pairs}}(f) \equiv \{N(\nu):N\{V_\beta\} \longrightarrow N\{U_\alpha\}\} \in \text{Pro}-H_{0,\text{Pairs}}.$$

Remark: The definition of $C_{0,\text{Pairs}}$ is modeled upon an analogous definition of $E_{0,\text{Pairs}}$ given by E. Friedlander [15]. The functor $C_{0,\text{Pairs}}$ seems to be new to shape theory. We shall see that it is very useful.

On page 20 we mentioned Mardesics' approach to shape which agrees with the Cech approach on paracompact spaces and agrees with the Cech approach based upon numerable coverings on all spaces. On page 15 we defined the Verdier functor which defines another shape theory on locally connected spaces and asked whether $\pi = C$ on locally connected metric spaces. If one uses only finite coverings, then one obtains a Cech theory C_f which satisfies $C_f(X) \simeq C(X^{\beta})$, where X^{β} is the Stone-Cech compactification of X.

Originally, Borsuk [5] and Fox [14] approached shape as follows. Let X be a compactum (= compact metric space) embedded in an ANR Q, for example, the Hilbert cube $I^{\infty} = \pi I_i$, $I_i = [-1,+1]$. Associate to $(X \xrightarrow{i} Q)$ the inverse system $\{U_{\alpha}\}$ of open neighborhoods of X in Q.

Def. IV.1: The intrinsic shape of a closed embedding $X \xrightarrow{i} Q$ of X into an ANR Q is

$$\text{INSH}(X \xrightarrow{i} Q) = \{U_{\alpha}\} \in \text{Pro-HANR},$$

where HANR is the homotopy category of ANRS.

Def. IV.2: The extrinsic shape of a closed embedding $X \xrightarrow{i} Q$ of X into an ANR Q is

$$\text{EXSH}(X \xrightarrow{i} Q) = \{U_{\alpha} - i(X)\} \in \text{Pro-HANR}.$$

It turns out that the isomorphism class of $\mathrm{INSH}(X \xrightarrow{i} Q)$ in Pro - HANR is independent of i and Q and depends only upon X. In particular, $\mathrm{INSH}(X \xrightarrow{i} Q)$ is canonically isomorphic to $C(X)$ in Pro - Htp, where Htp is the homotopy category of all topological spaces. On the other hand, the extrinsic shape is an important invariant of the embedding. For example, let $i:S^1 \longrightarrow S^3$ be a wild knot. There is a natural map

$$\mathrm{EXSH}(X^1 \xrightarrow{i} S^3) \xrightarrow{\phi} (S^3 - i(S^1)).$$

Taking fundamental groups, we obtain a map

$$\mathrm{Pro} - \pi_1(\mathrm{EXSH}(S^1 \xrightarrow{i} S^3)) \xrightarrow{\phi_*} \pi_1(S^3 - i(S^1)).$$

$\pi_1(S^3 - i(S^1))$ is the group of the knot i and the image of ϕ_* consists precisely of the peripheral elements of $\pi_1(S^3 - i(S^1))$. (See Fox [13] for a quick trip through knot theory).

The various constructions we have considered so far all yield functors to Pro - $\mathcal{H}$. Let CW be the category of CW - complexes and continuous maps and let SS be the category of simplicial sets and simplicial maps. The Cèch construction fails to define a functor to Pro - CW because refining maps are only unique up to homotopy, i.e., $\{\mathcal{N}\{U_\alpha\}\}$ is an inverse system in $\mathcal{H}$ but not in CW. On the other hand, the Vietoris construction defines an inverse system $\{\mathcal{V}\{U_\alpha\}\}$ in Pro - CW or in Pro - SS.

Def. IV.3: Let X be a topological space and $\{U_\alpha\}$ an open covering of X. Define $V\{U_\alpha\}$ to be the simplicial set whose n-simplices are ordered n+1-tuples $(x_0,\cdots,x_n)$ such that $\{x_0,\cdots,x_n\} \in U_\alpha$, for some member U_α, of the covering $\{U_\alpha\}$. If $\{\overline{U}_\beta\}$ refines $\{U_\alpha\}$, then we have a canonical inclusion $V\{\overline{U}_\beta\} \longrightarrow V\{U_\alpha\}$. Hence, $\{V\{U_\alpha\}\} \in$ Pro-SS and $\{|V\{U_\alpha\}|\} \in$ Pro-CW. We thus obtain a functor

$$\text{Top} \xrightarrow{V} \text{Pro-SS} \xrightarrow{|\cdot|} \text{Pro-CW}.$$

Remark: Having $V(X)$ defined in Pro-SS instead of in Pro-K is quite useful, e.g., one can take inverse limits in SS but not in K. On the other hand, $N\{U_\alpha\}$ is usually quite comprehensible while $V\{U_\alpha\}$ is almost always incomprehensible.

Dowker [40] has constructed a canonical homotopy equivalence $V\{U_\alpha\} \longrightarrow N\{U_\alpha\}$. Hence, the composition $\text{Top} \xrightarrow{V} \text{Pro-SS} \xrightarrow{|\cdot|} \text{Pro-CW} \longrightarrow \text{Pro-}H$ is canonically equivalent to $C:\text{Top} \longrightarrow \text{Pro-}H$.

For $(X \xrightarrow{f} Y) \in H_{0,\text{Pairs}}$ we have long exact sequences

$$\cdots \longrightarrow \pi_n(X) \longrightarrow \pi_n(Y) \longrightarrow \pi_n(f) \longrightarrow \pi_{n-1}(X) \longrightarrow \cdots,$$

$$\cdots \longrightarrow H_n(X) \longrightarrow H_n(Y) \longrightarrow H_n(f) \longrightarrow H_{n-1}(X) \longrightarrow \cdots,$$

$$\cdots \longrightarrow H^n(f) \longrightarrow H^n(Y) \longrightarrow H^n(X) \longrightarrow H^{n+1}(f) \longrightarrow \cdots.$$

Hence, for $\{X_\alpha \xrightarrow{f_\alpha} Y_\alpha\} \in \text{Pro}-\mathcal{H}_{0,\text{Pairs}}$ we have long exact sequences.

$$\text{IV.2.}\quad \cdots \longrightarrow \text{Pro}-\pi_n(X) \longrightarrow \text{Pro}-\pi_n(Y) \longrightarrow \text{Pro}-\pi_n(f) \longrightarrow \cdots,$$

$$\cdots \longrightarrow \text{Pro}-H_n(X) \longrightarrow \text{Pro}-H_n(Y) \longrightarrow \text{Pro}-H_n(f) \longrightarrow \cdots,$$

$$\cdots \longrightarrow H^n(f) \longrightarrow H^n(Y) \longrightarrow H^n(X) \longrightarrow \cdots,$$

where we have taken the direct limit, as usual, in the case of cohomology H^*. If we also take inverse limits, then we get long sequences

$$\cdots \longrightarrow \pi_n(X) \longrightarrow \pi_n(X) \longrightarrow \pi_n(f) \longrightarrow \cdots$$

$$\cdots \longrightarrow H_n(X) \longrightarrow H_n(X) \longrightarrow H_n(f) \longrightarrow \cdots,$$

which are not in general exact. The inverse limit functor is exact on Mittag-Leffler Pro-groups. (A Pro-group $\{G_\alpha\}$ is said to be Mittag-Leffler if it is isomorphic to a Pro-group $\{H_\beta\}$ all of whose bonding maps are surjections.)

Borsuk [5] introduced the notion of a movable space in order to guarantee that taking inverse limits wouldn't hurt.

Def. IV.4: Let X be a closed subset of a compact ANR Q. X is said to be movable if for every open neighborhood U of X in Q there exists an open neighborhood V, $X \subset V \subset U$, such that for all open neighborhoods W, $X \subset W \subset V \subset U$, V may be deformed into W within U.

Def. IV.5: Let C be a category and $X = \{X_\alpha\} \in \text{Pro}-C$. X is said to be movable if for every α there exists a $\beta > \alpha$ such that for all $\gamma > \beta$ there exists a map $X_\beta \xrightarrow{f} X_\gamma$ in C such that the diagram

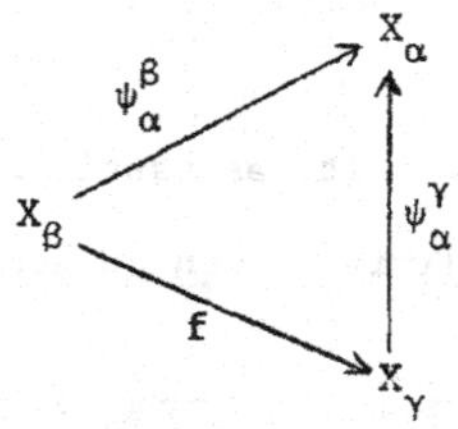

commutes, where ψ_α^β and ψ_α^γ are bonding maps of the inverse system $\{X_\alpha\}$ (which for simplicity we have assumed to be indexed by a directed set).

Remark: A movable Pro - group is Mittag-Leffler but the converse does not necessarily hold (consider $\{z_2 \longleftarrow z_{2^2} \longleftarrow z_{2^3} \cdots\})$.

Def. IV.6: A topological space X is said to be movable if $C(X)$ is movable in Pro $-H$. A map $(X \xrightarrow{f} Y) \in \text{Top}_{0,\text{Pairs}}$ is said to be movable if $C(f)$ is movable in $\text{Pro} - H_{0,\text{Pairs}}$.

Remark: Def. IV.4 and Def. VI.6 agree for compact metric spaces. Clearly every space having the shape of a CW - complex is movable.

Theorem IV.1 [44]: Cech homology and homotopy are exact on movable maps.

Theorem IV.2 [45]: The Cech Hurewicz Theorem holds on movable spaces.

Theorem IV.3 [32]: The Cech Whitehead Theorem holds on movable compacta having finite homotopy dimension.

Theorem IV.1 is proved by taking the inverse limit of IV.2 on p. 28 and observing that $\varprojlim$ is an exact functor on movable Pro-groups. Theorem IV.2 is proved by taking inverse limits in Theorem II.3, p. 11 and observing that a movable Pro-group whose inverse limit group is isomorphic to the zero group is itself isomorphic to the zero group in Pro-groups. Theorem IV.3 is proved by taking inverse limits in Theorem II.2, p.11 and using a non-trivial result of Moszynska [32] which allows one to pass from an isomorphism of the inverse limit homotopy groups to an isomorphism of the Pro-homotopy groups for movable compacta of finite homotopy dimension.

The following are some theorems characterizing movable spaces.

Theorem IV.4 (Borsuk [6]): Every planar compactum is movable. A product of a finite or a countable collection of movable compacta is again movable. The suspension of a movable compactum is movable.

Theorem IV.5 (Mardesic [28]): Every n-dimensional LC^{n-1} continuum is movable.

Note: X is LC^n if for each $x \in X$ and open neighborhood U of x there exist an open neighborhood V of x, $V \subset U$, such that the image $\pi_i(V) \longrightarrow \pi_i(U)$ is trivial for $i \leq n$.

Remark: Borsuk [4] has shown that every n-dimensional LC^n compactum is an ANR. Thus, movable spaces form a class of spaces more general than those having the shape of a CW-complex but still possessing many of the desirable properties of CW-complexes. In [11] the author and Ross Geoghegan have proved the following theorem using techniques of [8].

Theorem IV.6 (Edwards and Geoghegan [11]): Let X be a pointed topological space. There exists a pointed CW-complex Q, a map $q:Q \longrightarrow \mathcal{V}(X)$ and a spectral sequence $\{E_r^{p,q}(X)\}$ such that $E_2^{p,q}(X) = \underleftarrow{\mathrm{Lim}}^p \pi_q(\mathcal{V}(X))$ for $0 \leq p \leq q$, where $\mathcal{V}(X)$ Pro-CW is the Vietoris homotopy type of X. If X is movable, $\{E_r^{p,q}(X)\}$ converges completely to $\pi_*(Q)$. If X is a movable compactum, then $q_*:\pi_i(Q) \xrightarrow{\simeq} \pi_i(\ (X)) = \underleftarrow{\mathrm{Lim}}\ \pi_i(\mathcal{V}(X))$, and more generally $q:[W,Q] \xrightarrow{\simeq} \underleftarrow{\mathrm{Lim}}\ [W,\mathcal{V}(X)]$ for every pointed CW-complex W.

<u>Remark</u>: There is a natural functor R from Pro- $\mathcal{H}_0$ into the functor category $(\text{Sets})^{\mathcal{H}_0^0}$ defined by sending $\{X_\alpha\} \in \text{Pro}-\mathcal{H}_0$ to the functor $\varprojlim [—,X_\alpha]$. Theorem IV.6 implies that if X is a pointed, connected, movable compactum, then RC(X) is isomorphic in $(\text{Sets})^{\mathcal{H}_0^0}$ to a CW - complex Q. Under such circumstances we say that X has the very weak shape of a CW - complex. Hence,

<u>Theorem IV.7 ([11])</u>: Every connected, pointed movable compactum has the very weak shape of a CW - complex.

Sullivan [38] has proved a similar sounding result.

<u>Theorem IV.8 (Sullivan [38])</u>: If C is the class of finite groups and $X \in \text{Pro}-C\mathcal{H}_0$, then X has the very weak shape of a CW - complex.

<u>Warning</u>: A very weak shape equivalence $q:Q \longrightarrow X$ induces isomorphisms on inverse limit homotopy groups, but <u>NOT</u>, in general, on homotopy and homology Pro - groups or on inverse limit homology groups, or on direct limit cohomology groups.

One can ask for conditions which gurarantee that $q:Q \longrightarrow X$ is a weak shape equivalence, i.e., induces an isomorphism on homotopy Pro-groups, which by Artin and Mazur [3] is equivalent to being a $\natural$-isomorphism, which implies inducing isomorphisms on homology Pro-groups and on direct limit cohomology groups.

Theorem IV.9 (Edwards and Geoghegan [11]): Let X be a connected, pointed movable compactum whose inverse limit Cech homotopy groups, $\varprojlim \pi_i C(X)$, are discrete in the inverse limit topology. Then X has the weak shape of a CW-complex.

This theorem follows from Theorem IV.7 by using an observation of Atiyah and Segal's [2] that for Mittag-Leffler Pro-groups $\{G_\alpha\}$ the topologized inverse limit $\varprojlim G_\alpha$ completely determines the Pro-group $\{G_\alpha\}$.

Conjecture IV.1: Let X be a connected, pointed, movable compactum whose inverse limit Cech homotopy group $\varprojlim \pi_* C(X)$ is discrete in the inverse limit topology, where $\pi_*(W) = \prod_{i>0} \pi_i(W)$. Then X has the shape of a CW-complex.

Remark: Discreteness of $\varprojlim \pi_* C(X)$ is certainly necessary.

The following examples from [11] show how precise the above theorems are.

Example IV.1: Let T^∞ be the countably infinite product of circles S^1. T^∞ is movable, but $\check{\pi}_1(T^\infty)$ is not discrete. Hence, T^∞ has the very weak shape but not the weak shape of a CW - complex.

Example IV.2: Let $T_\infty^\infty = \prod_{n>0} S^n$ where S^n is the n - sphere. T_∞^∞ is movable and $\check{\pi}_i(T_\infty^\infty)$ is discrete for all i. But $\check{\pi}_*(T_\infty^\infty)$ is not discrete. Hence, T_∞^∞ has the weak shape but not the shape of a CW - complex.

Example IV.3: Let X_n be the wedge of spheres $\bigvee_{k\geq n} S^k$. Let $S^\infty = \{X_n\} = \{\bigvee_{k\geq n} S^k\} \in \text{Pro}-H_0$. Clearly, $\text{Pro}-\pi_i(S^\infty) \simeq 0$ for all i. Hence, S^∞ is weak shape equivalent to a point. But $\text{Pro}-\pi_*(S^\infty)$ is not movable, and hence S^∞ is not movable. Thus movability is not preserved under weak shape equivalenc.

Example IV.4: There is also a non-movable compactum X which is weak shape equivalent to a point. D. Kahn [47] (see also [20]) constructs X as the inverse limit of a system $X_0 \xleftarrow{\alpha_1} X_1 \xleftarrow{\alpha_2} X_2 \cdots$ with the following properties:

1. X_0 is a 7 - dimensional, 5 - connected, finite complex.
2. $X_{n+1} = \Sigma^4 X_n$, the 4 - fold suspension of X_n, and $\alpha_{n+1} = \Sigma^4 \alpha_n$ for all $n \geq 0$.

3. There is a map $\alpha : X_0 \longrightarrow S^3$ such that the compositions $f_n = \alpha \ \alpha_1 \cdots \alpha_n : X_n \longrightarrow S^3$ and all suspensions of the f_n are essential for $n \geq 1$.

Example IV.5: Let $X = \varprojlim \{S^2\}$ where the bonding maps are of degree 3. Then, (Sullivan [38], p. 3.4) X does not have the very weak shape of any CW - complex in the sense that there does not exist a CW - complex Q and a shape morphism $Q \xrightarrow{q} X$ such that $[-,Q] \xrightarrow{q_*} \varprojlim [-, C(X)]$ is a natural equivalence.

Now let X and Y be compacta contained in the pseudo-interior S of the Hilbert cube I^∞, where $S = \prod_{i>0} I_i^0 = (-1,+1)$, $T^\infty = \prod_{i>0} I_i$, $I_i = [-1,+1]$.

Theorem IV.10 (Chapman [9]): X and Y have the same shape iff their complements $I^\infty \setminus X$ and $I^\infty \setminus Y$ are homeomorphic.

Theorem IV.11 (Geoghegan and Summerhill ([17]): Let X and Y be non-empty compact strong Z_{n-k-2} - sets in R^n $(k \geq 0, n \geq 2k+2)$. Then the following are equivalent:

1. $\mathrm{InSh}(X) \simeq \mathrm{InSh}(Y)$;
2. $(R^n/X, X/X)$ and $(R^n/Y, Y/Y)$ are homeomorphic as pairs;

3. $R^n - X$ and $R^b - Y$ are homeomorphic.

Remark: This theorem applies in particular to the case of tamely embedded k-dimensional polyhedra in R^{2k+2}. We conjecture that one can drop the tameness assumption if one replaces INSH by EXSH (see Def. IV.1 and Def. IV.2).

Mardesic and Segal [31] have classified solenoids and sphere like continua up to shape. Consider the inverse system $\underline{S}_P = \{X_n, \pi_{n,n+1}\}$, where $X_n = \{z \mid |z| = 1\}$ is the unit circle in the complex plane and the map $\pi_{n,n+1} : X_{n+1} \longrightarrow X_n$ is given by $\pi_{n,n+1}(z) = z^{p_n}$ where $P = (p_1, p_2, \cdots)$ is a sequence of primes. Let $S_P = \underleftarrow{\mathrm{Lim}}\, \underline{S}_P$. Two sequences of primes $P = (p_1, p_2, \cdots)$ and $Q = (q_1, q_2, \cdots)$ are said to be equivalent, written $P \sim Q$, provided it is possible to delete a finite number of terms from each so that every prime occurs the same number of times in each of the deleted sequences.

Theorem IV.12 (Mardesic and Segal [31]): Let S_P and S_Q be two solenoids. Then, the following three statements are equivalent:

1. S_P and S_Q are of the same shape;

2. $P \sim Q$;

3. S_P and S_Q are homeomorphic.

Keesling [23] has proved the following generalizations of Theorem IV.12.

Theorem IV.13 (Keesling [23]): Two compact, connected abelian topological groups have the same shape iff they are isomorphic.

Theorem IV.14 ([23]): Let G be a compact connected topological group and A a compact connected abelian topological group. Then any shape morphism $F:\mathcal{C}(G) \longrightarrow \mathcal{C}(A)$ is determined by a unique continuous homomorphism $f:G \longrightarrow A$.

Remarks: Keesling's results suggest that the notion of shape may become a useful tool in the theory of topological groups. For example, every compact connected abelian topological group G has the shape of a Pro - Eilenberg - MacLane space, namely of its Lie Series $L(G)$ which is an inverse system of Tori. Let LG be the category of Lie Groups and Lie Group Homomorphisms. If G is a topological group, then $(G \downarrow LG)$ is filtering. This is because the product of two Lie Groups is a Lie Group and a closed subgroup of a Lie Group is a Lie Group. We can thus associate to any topological group G its fundamental Lie Series $(G \downarrow LG) \longrightarrow LG$, and hence obtain a functor $L:\mathrm{Top}\, G \longrightarrow \mathrm{Pro} - LG$. If $\{A_n\}$ is an inverse system of compact connected abelian Lie Groups (i.e., Tori) and $A = \varprojlim A_n$, then $L(A) \simeq \{A_n\}$ in Pro - LG. In other words, any two Lie Series for A are isomorphic in

Pro - $\mathcal{LG}$. This is shown by using Theorem 11.9, p. 287 of Eilenberg and Steenrod [41] and Scheffer [42]. The situation for non-abelian groups seems more difficult and one will probably have to pass to a homotopy theory, possibly involving A_∞ - spaces and A_∞ - maps [43] [12].

Theorem IV.12 generalizes to

Theorem IV.15 ([31]): Two spaces S_P^m and S_Q^m are of the same shape if and only if $P \sim Q$.

Definition IV.7: A metric continuum X is said to be S^m - like provided for each $\epsilon > 0$ there is a mapping $f_\epsilon : X \longrightarrow S^m$ onto S^m such that DIAM $f^{-1}(y) < \epsilon$ for any $y \in S^m$.

Theorem IV.16 ([31]): Every S^m - like continuum X has the shape of a point, X^m or S_P^m.

V. Classification

This section is concerned with the problem of extending the usual classification theorems of algebraic topology from $\mathcal{H}_0$ to Pro - $\mathcal{H}_0$ and to Top_0. We will begin with the case of covering spaces.

Def. V.1: $E \xrightarrow{p} X$ is a covering space if every $x \in X$ has an open neighborhood U such that $p^{-1}(U)$ is a disjoint union of open set S_i in E, each of which is mapped homeomorphically onto U by p. Such U are said to be evenly covered, and the S_i are called sheets over U. Two covering spaces $(E_i,e_i) \xrightarrow{p_i} (X,x_0)$, $i = 1,2$, are equivalent if there is a homeomorphism $\phi:(E_1,e_1) \longrightarrow (E_2,e_2)$ such that $p_2\phi = p_1$.

Theorem V.1 (The Fundamental Theorem of Covering Spaces): Let X be a CW-complex and H a subgroup of $\pi_1(X,x_0)$. Then, there exists a covering space $(E_H,e_0) \xrightarrow{p} (X,x_0)$, unique up to equivalence, such that $H = p_*\pi_1(E_H,e_0)$. Hence, equivalence classes of covering spaces are in bijective correspondence with subgroups of $\pi_1(X,x_0)$.

Remark: If we allow the equivalence ϕ not to preserve base points, then equivalence classes of covering spaces will only correspong to conjugacy classes of subgroups of $\pi_1(X,x_0)$. The fundamental theorem actually holds for X which are connected, locally path connected, and semi-locally 1-connected, but fails to hold as stated for more pathological spaces e.g., the Warsaw circle.

Artin and Mazur [3] have given the following characterization of E_H.

<u>Theorem V.2 ([3])</u>: Let X be a CW-complex and H a subgroup of $\pi_1(X,x_0)$. Let $E_H \longrightarrow X$ be the covering space corresponding to H by the fundamental theorem of covering spaces. Then $E_H \longrightarrow X$ is characterized by:

For any CW-complex W the map $E_H \longrightarrow X$ identifies $[W, E_H]$ with the subset of $[W,X]$ of maps which carry $\pi_1(W)$ into the subgroup H of $\pi_1(X)$.

Let $X = \{X_j\} \in \text{Pro}-\mathcal{H}_0$, and let $H \hookrightarrow \text{Pro}-\pi_1(X)$ be a sub-Pro-Group.

<u>Theorem V.3 ([3])</u>: A map $f:X \longrightarrow Y$ of Pro-objects of a category $\mathcal{C}$ can be represented, <u>up to isomorphism</u>, by a filtering inverse system of maps $\{f_i':X_i' \longrightarrow Y_i'\}$, i.e., by a Pro-object in the category of maps of $\mathcal{C}$. If $\mathcal{C}$ is an abelian category, then a monomorphism (epimorphism) in $\text{Pro}-\mathcal{C}$ can be represented by an inverse system of monomorphisms (epimorphisms) in $\mathcal{C}$. Similarly for $\mathcal{C}$ = the cateogry of groups.

Hence, we may assume that H and X have the same indexing category and that we are given compatible injections $H_j \hookrightarrow \pi_1(X_j)$ representing the map $H \longrightarrow \text{Pro}-\pi_1(X)$. From the fundamental theorem of covering spaces we obtain an inverse system of covering spaces $(E_j)_{H_j} \longrightarrow X_j$, and hence a map of Pro-objects $\{(E_j)_{H_j}\} = E_H \longrightarrow X = \{X_j\}$.

<u>Theorem V.4 (Fundamental Theorem of Covering Spaces in Pro $-\mathcal{H}_0$)</u>: Let $X \in \text{Pro}-\mathcal{H}_0$, and let $H \hookrightarrow \text{Pro}-\pi_1(X)$ be a monomorphism of Pro-groups. There is an $E_H \in \text{Pro}-\mathcal{H}_0$ together with a map $E_H \longrightarrow X$ which is characterized by the property that for each $W \in \text{Pro}-\mathcal{H}_0$, $[W, E_H]$ is carried to the subset of $[W,X]$ of maps such that the induced map $\text{Pro}-\pi_1(W) \longrightarrow \text{Pro}-\pi_1(X)$ factors through H. E_H is called the covering space of X determined by the sub-Pro-Group $H \hookrightarrow \text{Pro}-\pi_1(X)$.

Thus, the fundamental theorem of covering spaces in $\mathcal{H}_0$ extends nicely to $\text{Pro}-\mathcal{H}_0$. The situation in Top_0 is more complicated. If $X \in \text{Top}_0$ and $H \hookrightarrow \text{Pro}-\pi_1(C(X))$, then any covering space of X corresponding to H, $E_H \longrightarrow X$, (assuming one exists) should certainly satisfy the condition that $C(E_H) \longrightarrow C(X)$ is an Artin-Mazur covering space as described in Theorem V.4. Any such map $E_H \longrightarrow X$ will be called a realization of $C(X)_H \longrightarrow C(X)$ in Top_0. We are now faced with two problems:

<u>Problem 1</u>: Define a notion of covering space in Top_0 such that its shape is a covering space in $\text{Pro}-\mathcal{H}_0$.

<u>Problem 2</u>: Show that every covering space in $\text{Pro}-\mathcal{H}_0$ over $C(X)$, $X \in \text{Top}_0$, can be realized by a map in Top_0 which is a covering space in Top_0.

Remarks: We can, of course, simply call any map $E_H \xrightarrow{f} X$ in Top_0 a shape covering space if $C(E_H) \xrightarrow{C(f)} C(X)$ is a covering space in $Pro-H_0$. This sidesteps Problem 1 but still leaves Problem 2. The analogous problems for schemes is answered by

Theorem V.5 ([15]): Let X be a pointed connected Northerian Scheme and $H \hookrightarrow Pro-\pi_1(X_{et})$ a sub-Pro-Group of $Pro-\pi_1(X_{et})$. Then there is a pointed étale cover $X_H \longrightarrow X$, such that $(X_H)_{et} \longrightarrow X_{et}$ is an Artin-Mazur covering space corresponding to H.

Fox [14] has introduced the following refinement of the usual notion of covering space.

Def. V.2: Let $\tilde{X} \xrightarrow{p} X$ be a map in Top_0. A collection $\tilde{M} = \{\tilde{M}_i^\alpha\}$ of subsets of $\tilde{X}$ will be said to lie evenly over a collection $M = \{M_i\}$ of subsets of X when: $p^{-1}(M_i) = {}_\alpha \tilde{M}_i^\alpha$ for each index i; each $\tilde{M}_i^\alpha$ is open in $p^{-1}(M_i)$; each set $\tilde{M}_i^\alpha$ is mapped by $p_i^\alpha = p|\tilde{M}_i^\alpha$ topologically onto M_i; and if $M_i \cap M_j \neq \phi$, then each set $\tilde{M}_i^\alpha$ meets exactly one of the sets $\tilde{M}_j^\beta$ (in particular, $M_i^\alpha \cap M_i^\beta = \phi$ whenever $\alpha \neq \beta$). $p:\tilde{X} \longrightarrow X$ will be called an overlay if $\tilde{X}$ has an open cover $\tilde{M}$ that lies evenly over some open cover M of X.

<u>Remark</u>: If $\tilde{X} \xrightarrow{p} X$ is a finite overlay, then the Cech type of p, $C(p) = \{N\{V_\beta\} \xrightarrow{Np_{\alpha\beta}} N\{U_\alpha\}\}$, has a cofinal subsystem of covering spaces, and hence $C(\tilde{X}) \xrightarrow{p_*} C(X)$ is a covering space in Pro-H_0. This is because if $\tilde{M}$ lies evenly over M, then $N(\tilde{M}) \longrightarrow N(M)$ is a covering space, and we can find a cofinal family $(\tilde{N},N,\nu)$ of coverings of p such that $\tilde{N}$ lies evenly over N. If $(\tilde{N},N,\nu)$ refines $(\tilde{M},M,\mu)$, then the covering space $N(\tilde{N}) \xrightarrow{N(\nu)} N(N)$ is equivalent to the pullback of $N(\tilde{M}) \xrightarrow{N(\mu)} N(M)$ via the refining map $N(N) \longrightarrow N(M)$. Hence, the system $\{N(\tilde{N}) \longrightarrow N(N)\}$ is of a very special and simple type. If $\tilde{X} \xrightarrow{p} X$ is an infinite overlay, then the system $\{N(\tilde{N}) \longrightarrow N(N)\}$ is still of pullback type, but is no longer cofinal in $C(p)$. Thus, $\{N(\tilde{N}) \longrightarrow N(N)\}$ is always a covering space in Pro-H_0 of pullback type, but $C(p)$ may fail to be one (consider the universal covering space of the one point compactification of $S^1 \vee S^2 \vee S^2 \vee \cdots$). Let O be the category of overlays and HCS the homotopy category of CW-covering

spaces. We have a functor $F: \mathcal{O} \longrightarrow$ Pro HCS which sends $(\tilde{X} \xrightarrow{p} X)$ to $\{N(\tilde{N}) \longrightarrow N(N)\}$. Thus, Problem 1 is solved by Fox by redefining <u>both</u> the notion of covering space and the notion of the shape of a covering space!

<u>Theorem V.6 (Fox [14])</u>: (Fundamental Theorem of Overlay Theory) The d - fold overlays $(\tilde{X},\tilde{x})$ of a pointed connected paracompact space (X,x) are in biunique correspondence with equivalence classes of representations in the symmetric group Σ_d of degree d of the Pro - group Pro - π_1 $(C(X,x))$.

<u>Remark:</u> One can obtain a generalization of Theorem V.6 to all spaces using numerable coverings and numerable overlays. Note that Fox does <u>not</u> say that to every sub - Pro - group H of Pro - $\pi_1(X,x)$ there corresponds an overlay $X_H \longrightarrow X$. The following is an example of a shape covering space which is not equivalent to any overlay.

<u>Example V1</u>: Let $D_2 = \{X_n\}$, where $X_n = \{z \in \mathbb{C} \mid |x| = 1\}$ and $p_{n,n+1}(z) = z^2$, and let $D_{2,3} = \{X_n\}$, with $p_{n,n+1}(z) = z^6$. The map $f: D_{2,3} \longrightarrow D_2$ given by $f_n: X_n \longrightarrow X_n$ which sends $z \longrightarrow f(z) = z^{(3^n)}$, is the covering space in Pro - H_0 corresponding to the sub - Pro - group Pro - $\pi_1(D_{2,3}) \hookrightarrow$ Pro - $\pi_1(D_2)$. Let $\underline{D}_2 = \varprojlim D_2$ and $\underline{D}_{2,3} = \varprojlim D_{2,3}$ and $\underline{f} = \varprojlim f$. The

sequence $0 \longrightarrow \hat{Z}_3 \longrightarrow \underline{D}_{2,3} \longrightarrow \underline{D}_2 \longrightarrow 0$ is a short exact sequence of compact abelian topological groups, where $\hat{Z}_3$ is the 3 - adic integers. $\underline{f}$ is clearly not equivalent to any overlay since $f_*(\text{Pro}-\pi_1(D_{2,3}))$ is not the kernel of any homomorphism from $\text{Pro}-\pi_1(D_2)$ to <u>any</u> group. The following diagram may help.

$$\begin{array}{ccccccccc}
0 & \longrightarrow & \hat{Z}_3 & \longrightarrow & \underline{D}_{2,3} & \xrightarrow{f} & \underline{D}_2 & \longrightarrow & 0 \\
 & & \downarrow & & \downarrow & & \downarrow & & \\
 & & \vdots & & \vdots & & \vdots & & \\
 & & \downarrow & & \downarrow & & \downarrow & & \\
0 & \longrightarrow & Z_{3^n} & \longrightarrow & S^1 & \xrightarrow{3^n} & S^1 & \longrightarrow & 0 \\
 & & \downarrow & & \downarrow{\scriptstyle 2\cdot 3} & & \downarrow{\scriptstyle 2} & & \\
0 & \longrightarrow & Z_{3^{n-1}} & \longrightarrow & S^1 & \xrightarrow{3^{n-1}} & S^1 & \longrightarrow & 0 \\
 & & \downarrow & & \downarrow & & \downarrow & & \\
 & & \vdots & & \vdots & & \vdots & &
\end{array}$$

Note that $f:D_{2,3} \longrightarrow D_2$ is <u>not</u> of pullback type.

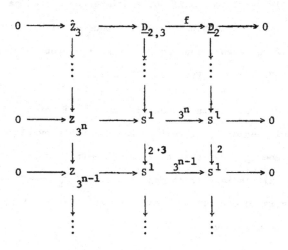

Let $CW_{0,f}$ be the category of pointed connected finite CW - complexes. If $\{\tilde{X}_n \xrightarrow{f_n} X_n\}$ is an inverse system of covering spaces in $CW_{0,f}$, then $\varprojlim \tilde{X}_n \xrightarrow{\varprojlim f_n} \varprojlim X_n$ is a light open fibration between compact spaces (light = totally disconnected inverse images of points). We shall call such maps profinite covering spaces.

Theorem V.7: Let $X \in \text{Pro}-CW_{0,f}$ and let $H \hookrightarrow \text{Pro}-\pi_1(X)$ have Pro-finite cokernel, i.e., $\{\pi_1(X_\alpha)/H_\alpha\}$ is isomorphic to a Pro-finite group. Then there is a profinite covering space $E \xrightarrow{p} Y$ such that: 1. $X \xrightarrow[\phi]{\simeq} C(Y)$ in $\text{Pro}-\mathcal{H}_0$;
2. (p) has a cofinal subsystem of finite covering spaces and thus determines a covering space in $\text{Pro}-\mathcal{H}_0$.
3. $\phi_* H = p_*(\text{Pro}-\pi_1(C(E)))$.

Theorem V.7 is about the best result one could have reasonably hoped for. The moral of the above is that one first tries to extend results from $\mathcal{H}_0$ to $\text{Pro}-\mathcal{H}_0$. This step should be easy. Next, one tries to realize in TOP a pro-structure in Pro-CW. For structures in $\text{Pro}-CW_f$ we can just take inverse limits, but the inverse limit structures we obtain may be more general than we expected. In [12] we will consider in detail the above program for principal bundles and fibrations.

REFERENCES

[1] G. Allaud, On the classification of fiber spaces, Math. Zeitschr. 92, 110-125 (1966).

[2] M. F. Atiyah and B. G. Segal, Equivariant K-theory and Completion, J. Diff. Geo. 3 (1969), 1-18.

[3] M. Artin and B. Mazur, Etale Homotopy, Lecture notes in Math. 100. Springer (1969).

[4] K. Borsuk, Theory of Retracts, Monografie Matematyczne 44, Warzawa 1967.

[5] ________, Concerning Homotopy Properties of Compacta, Fund. Math. 62 (1968), 223-254.

[6] ________, On movable compacta, Fund. Math. 66 (1969), 137-146.

[7] ________, Theory of Shape, Lecture Notes No. 28, Matematisk Inst., Aarhus Univ., 1971, 1-145.

[8] A. K. Bousfield and D. M. Kan, Homotopy Limits, Completions and Localizations, Lecture notes in Math. 302 (1972).

[9] T. A. Chapman, On some applications of infinite dimensional manifolds to the theory of shape, (to appear).

[10] J. Cohen, The homotopy groups of inverse limits, Proceedings of the London Math. So., Vol. XXVII, July, 1973, 159-192.

[11] D. A. Edwards and R. Geoghegan, Compacta weak shape equivalent to ANR's, (to appear in Fund. Math.)

[12] ________ and P. McAuley, Pro-fibrations, shape and classification, (in preparation).

[13] R. H. Fox, A quick trip through Knot Theory, in Topology of 3-Manifolds, ed. M. K. Fort, Jr., Prentice-Hall Inc., Englewood Cliffs, N. J., 1962.

[14] ________, On Shape, Fund. Math. LXXIV (1972), 47-71.

[15] E. Friedlander, Fibrations in Etale Homotopy Theory, to appear.

[16] ________, The Etale Homotopy Theory of a Geometric Fibration, to appear.

[17] R. Geoghegan and R. Summerhill, Concerning the shapes of finite dimensional compacta, Trans. Amer. Math. Soc. 179 (1973), 281-292.

[18] M. Greenberg, Lectures on Algebraic Topology, Benjamin, Amsterdam, 1967.

[19] A. Grothendieck, Technique de descente et theorems d'existence en geometrie algebrique II, Seminar Bourbaki, 12 iene annee, 1959-60, exp. 195.

[20] D. Handel and J. Segal, An acyclic continuum with non-movable suspensions, (to appear).

[21] ________, Finite Shape Classifications, (to appear).

[22] D. Husemoller, Fibre Bundles, McGraw-Hill Book Company, 1966.

[23] J. Keesling, Continuous Functions induced by Shape Morphisms, to appear.

[24] D. Knutson, Algebraic Spaces, Lecture notes in Math. 203, Springer (1971).

[25] J. Levan, Shape Theory, Thesis 1973 (unpublished).

[26] A Lundell and S. Weingram, The topology of CW complexes, Van Nostrand Reinhold Company (1969).

[27] I. G. MacDonald, Algebraic Geometry - Introduction to Schemes, W. A. Benjamin, Inc., 1968.

[28] S. Mardesic, n-dimensional LC^{n-1} compacta are movable, Bull. Acad. Polon. Sci. Sec. Sci. Math. Astr. Phys. (1971), 505-509.

[29] ________, Shapes for topological spaces, to appear.

[30] ________, A survey of the shape theory of compacta, to appear.

[31] ________ and J. Segal, Shapes of compacta and ANR-systems, Fund. Math. LXXII, (1971), 41-59.

[32] M. Moszynska, The Whitehead Theorem in the Theory of Shapes, to appear in Fund. Math.

[33] L. Pontrjagin, Topological Groups, Princeton University Press, 1946.

[34] T. Porter, Cech Homotopy I, J. London Math. Soc. (2), 6 (1973) 429-436.

[35] D. Quillen, Some remarks on étale homotopy theory and a conjecture of Adams, Topology 7 (1968) 111-116.

[36] J. P. Serre, Groupes d'homotopie et classes de groupes abeliene, Annals of Math., Vol. 58, (1953), 258-294.

[37] E. Spanier, Algebraic Topology, McGraw-Hill, 1967.

[38] D. Sullivan, Geometric Topology, Part I, Localization, Periodicity and Galois Symmetry, Mimeographed Notes M.I.T. 1970.

[39] A. Dold, Lectures on Algebraic Topology, Springer-Verlag, New York, 1972.

[40] C. H. Dowker, Homology of Relations, Ann. of Math, 56 (1952), 84-95.

[41] S. Eilenberg and N. Steenrod, Foundations of Algebraic Topology, Princeton University Press, Princeton, New Jersey (1952).

[42] W. Scheffer, Maps between topological groups that are homotopic to homomorphism, Proc. Amer. Math. Soc. 33 (1972), 562-567.

[43] J. Stasheff, H-spaces from a homotopy point of view, Lecture notes in Mathematics 161, Springer (1970).

[44] R. Overton, Cèch homology for movable compacta, to appear in Fund. Math.

[45] K. Kuperberg, An isomorphism theorem of Hurewicz type in Borsuk's theory of shape, to appear.

[46] N. Steenrod, The topology of fibre bundles, Princeton University Press (1951).

[47] D. S. Kahn, An example in Cèch cohomology, Proc. Amer. Math. Soc. 16 (1965), 584.

HOMOTOPY ASSOCIATIVE CATEGORIES

by

Pierre J. Malraison, Jr.

Introduction.

A recent development in algebraic topology is the construction of "non-categories" which fail to have associative compositions. The standard remedy is to pass to homotopy classes of maps, since the composition is usually homotopy associative. This corresponds to replacing a homotopy associative H - space by its path components, and ignores a lot of the structure available. The purpose of this note is to begin a theory of homotopy associative categories (including higher homotopies).

Section One sets up the structure in terms of operads. Section 2 compares the operad D with others in the literature. Section 3 translates the Kleisli construction to the context of homotopy associative categories. Section 4 gives some examples. Section 5 is an outline of work in progress, and contains some comments about what homotopy functors between homotopy associative categories ought to be.

I would like to thank the National Science Foundation for supporting part of this research under Grant No. GP-38595. I would also like to thank Carleton College and the Shell Foundation for Supplementing my travel funding.

Section 1. Operads.

An operad is the bare essential of a PROP or category of operators [MacLane, Boardman-Vogt]. Since the problems under consideration relate to associativity and not commutativity, no action by the symmetric groups is assumed. In the definition below P(0) is not restricted to being one point, but the particular system set up will have that property. An operad is a pre-operad in [May].

Definition 1. An operad is a sequence of space $P(n)$, $n \geq 0$, and maps

$$\mu(n_1,n_2,\cdots,n_k): \prod_{i=1}^{k} P(n_i) \times P(k) \longrightarrow P\left(\sum_{i=1}^{k} n_i\right)$$

which satisfy the following conditions:

i) (associativity)

$$\mu(f_1,f_2,\cdots,f_k,\ \mu(g_1,g_2,\cdots,g_m,h)) = \mu(w_1,w_2,\cdots,w_n,h)$$

where

$$k = \sum_{n=1}^{m} \dim (g_j),\ g_j \in P(\dim (g_j))$$

$$w_i = \left(f_{i_1}, f_{i_1+1}, f_{i_1+2}, \cdots, f_{i_1+\dim (g_i)-1}, g_i\right)$$

$$i_1 = \left(\sum_{j=1}^{i-1} \dim (g_j)\right) + 1 .$$

ii) (unitary) There is an $e \in P(1)$ such that

$$\mu(f,e) = f$$

$$\mu(e,e,\cdots,e,f) = f .$$

In the definition the indices of the μ's have been suppressed.

Definition 2. Let P and P' be operads. A morphism $\phi{:}p \longrightarrow P'$ is determined by a sequence of continuous maps:

$$\phi(n){:}\ \ P(n) \longrightarrow P'(n)$$

such that

$$\phi(m)(\mu(f_1,f_2,\cdots,f_k,g) = \mu'(\phi(i_1)(f_1),\phi(i_2)(f_2),\cdots,\phi(k)(g)).$$

Examples.

1. $A(n) = 1$, the one point space for all n. The μ's are all given by identity maps.

2. Let $\underline{X}$ be a topological category with the following data:

   i) an associative bifunctor $\otimes : \underline{X} \times \underline{X} \longrightarrow \underline{X}$

   ii) an identity 1 for $\otimes$, i.e., an object in $|\underline{X}|$ such that

$$1 \otimes X = X \otimes 1 = X .$$

Then let $X^{(n)} = X \otimes X^{(n-1)}$, $X^{(0)} = 1$, $X^{(p)} = X$.

The operad End (X) is defined by

$$\text{End }(X)(n) = \underline{X}(X^{(n)}, X),$$

$$\mu(f_1, f_2, \cdots, f_k, g) = g \cdot f_1 \otimes f_2 \otimes \cdots \otimes f_k.$$

The identity in $\underline{X}(X,X)$ is an identity for μ in the sense of Definition 1, ii. Strictly speaking, one should write End $(X, \underline{X}, \otimes)$ but in most cases the x will be either clear from the context or explicitly described.

<u>Definition 3.</u> Let P be an operad, X a topological category as in Example 2. Then an object X in $|X|$ is a P - object if there is a morphism of operads

$$d: P \longrightarrow \text{End }(X).$$

<u>Proposition 4.</u> An A — object in $(\underline{Top}, \otimes)$ is a topological monoid.

<u>Proof.</u> Immediate. The multiplication is the image of 1 in (X^2, X). Associativity then follows from the fact that $m \cdot (id \times m)$ and $m \cdot (m \times id.)$ both must be the image of 1 in (X^3, X). $X^0 = 1$ the one-point space, and a similar argument shows that the element picked out by the image of 1 in (X^0, X) is an identity for the multiplication. The details are left to the reader as an exercise in using

operads.

This section concludes with the definition of the operad that will define homotopy associative structures.

Definition 5. The operad D is defined as follows:

$$D(n) = \{f:[0,1] \longrightarrow [0,n] \mid f(0) = 0,\ f(1) = n,\ f \text{ is continuous}\}$$

$$\mu(n_1,n_2,\cdots,n_k)(f_1,\cdots,f_k,g)(t) = f_i(g(t) - i+1) + \sum_{j=1}^{i=1} n_j ,$$

for $i = 1,2,\cdots,k$ and $i-1 \leq g(t) \leq i$.

Remarks.

1. Well-definedness. The only problem is at points where $g(t) = i$. At such points the two definitions would be

$$f_i(1) + \sum_{j=1}^{i-1} n_j \quad \text{and} \quad f_{i+1}(0) + \sum_{j=1}^{i+1-1} n_j ,$$

which are equal, since $f_i(1) = n_i$ and $f_{i+1}(0) = 0$.

2. Low dimensional $D(i)$ and unitary property. $D(0) = 1$, the one point space. $D(1)$ includes the identity map $I \longrightarrow I$. Clearly

$$\mu(f,id.)(r) = f(id.(r)) = f(r),$$

so $\mu(f,id.) = f$.

$$\mu(\text{id.},\text{id.},\cdots,\text{id.},g)(t) = (g(t) - i + I) + \sum_{j-1}^{i-1} 1$$

$$= g(t) - i + 1 + (i - 1)$$

$$= g(t) .$$

So D satisfies Definition 1, ii.

3. Associativity. The calculation follows:

$$\mu(f_1,\cdots,f_k, \mu(g_1,\cdots,g_m.h))(t)$$

$$= f_i(\mu(g_1,\cdots,g_m,h)(t) - i + 1) + \sum_{j=1}^{i-1} \dim (f_j),$$

$$\text{for } i - 1 \leq \mu(g\cdots h)(t) \leq i$$

$$= f_i((g_q(h(t) - q + 1) + \sum_{p=1}^{q-1} \dim (g_p)) - i + 1) + \sum_{j=1}^{i=1} \dim (g_j)$$

$$\text{for } q - 1 \leq h(t) \leq q \text{ and}$$

$$i - 1 < \mu(g_1,\cdots,g_m,h)(t) \leq i$$

$$= f_i((g_q(h(t) - q + 1) + (s - i) + 1) + \sum \dim (f_j) \qquad s = \sum_{p=1}^{q-1} \dim (g_p)$$

$$= f_i((g_q(h(t) - q + 1) - (i - s) + 1)$$

$$+ \sum_{p=1}^{q-1} \dim\left(\mu\left(f_{i_p},\cdots,f_{i_p + \dim(q_p) - 1}, g_p\right)\right) + \sum_{u=i_p + \dim(g_p)}^{i-1} \dim(f_j)$$

$$= f_{i_p + (i-s)}((g_q(h(t) - q + 1) - (i - s) + 1) + \text{sum terms})$$

$$= \mu(w_1,w_2,\cdots,w_n,h) \text{ in the notation of Definition 1, i.}$$

The main difficulty above is that

$$i-1 \leq \mu\,(g_1,\cdots,g_m,h) \leq i$$

implies that

$$i-s-1 \leq g_q(h(t)-q+1) \leq i-s \quad \text{if} \quad g-1 \leq h(t) \leq q\,.$$

Section 2. Other Theories of Homotopy Associativity.

This section compares D to other theories of homotopy associativity. Specifically Stasheff's A_∞ and Boardman-Vogt's WA. For typographical convenience, and following [Stasheff], we shall write K(n) in place of $A_\infty^{(n)}$, and refer to the operads as K. In the paper cited above, K was not given all the formal structure of an operad. The structure given to K will be that induced by the map of Theorem 8.

Following the fourth of Boardman's options for "basepoint watchers," we will omit them. In the case of $K \longrightarrow WA$ this is necessary, since K assumes strict identities, while WA does not. In terms of operads, the requirement is $P(0) = \phi$ and $P(1) = 1$, the identity.

Definition 6. The spaces K(n) are defined as follows:

$K(2) = 1$, the one point space.

$K(n) = CL(n)$, the cone on $L(n)$ for $n > 2$, where

$$L(n) = \bigcup_{\substack{r+s=n+1 \\ r,s > 2}} K(r) \times K(s) \times \langle r \rangle/\sim, \text{ where}$$

$\langle r \rangle = \{1,2,3,\cdots,r\}$, and $\sim$ is generated by two sorts of relations:

a) $(x,(z,w,k),j) = ((x,z,j),w,j+k-1)$

b) $99x,z,k),w,j+s-1) = ((x,w,j),z,k)$.

<u>Remark</u>. In the relations a) and b) (z,w,k) represents an element of $L(r)$, considered as an element of $K(r)$. These relations are straightforward translations of relations 3 (a), (b) in [Stasheff, p. 278].

<u>Definition 7</u>. The operad WA [Boardman]. $WA(n)$ is the set of all trees with vertex labels in A, n twigs and one root. Each internal edge also has a length from 0 to 1. Identifications are made as follows:

a) Edges of length 0 may be collapsed.

b) Vertices labelled by the identity may be removed.

<u>Remark</u>. We refer the reader to [Boardman] for examples of the relations. μ is defined by

$\mu(\alpha_1,\alpha_2,\cdots,\alpha_k,\beta)$ = the tree obtained by grafting the tree α_i onto the i-th twig of β , and giving the new internal edge length 1.

<u>Theorem 8</u>. There is an injection for each n, $K(n) \longrightarrow WA(n)$. Using this injection to induce an operad structure on the $K(n)$, operad morphism $K \longrightarrow WA$ is obtained.

<u>Proof</u>. Define $f(1)$: $K(1) \longrightarrow WA(1) = 1 \to 2$ (the tree with two twigs, one root, vertex label the unique element of $A(2)$) if $i = 2$. Now suppose $F(n)$ is given for $n < m$. Define

$F(r,s,m):K(r) \times K(s) \times \langle r \rangle \longrightarrow WA(m)$ by

$$(x,y,j) \longmapsto \mu(id,id,\cdots,F(s)(y),\cdots,id,F(r)(x))$$

where $F(s)(y)$ appears in the j-th position. It is easy to show that the $F(r,s,m)$ glue together to give a map $L(m) \longrightarrow WA(m)$. To extend the map to $K(m)$, map the vertex of the cone to m, the tree with m twigs, one root, and vertex label the unique element of $A(m)$. Let $F(r,s,m,t)$ be the same as $F(r,s,m)$, except graft with length t instead of 1. Then $F(r,s,m,0) = m$, and glued together the $F(r,s,m,t)$ give a map on the cone.

To complete the proof it must be shown that when images of K are combined via μ the result is also an image of K. Repeated use of the associativity of μ yields:

$$\mu(f(s_1)(x_1), f(s_2)(x_2), \cdots, f(s_n)(x_n), f(n)(y))$$

$$= f \sum_{i=1}^{n} s_i + t) ((\cdots\cdot(y,x_1,1),x_2,2),x_3,3),\cdots\cdot),x_n,n)$$

Theorem 9. There is an operad morphism $WA \longrightarrow D$..

Proof. Filter $WA(n)$ by number of internal edges. $WA^i(n)$ is then homeomorphic to $W^i(n) \times I^i$ where $W(n)$ is all trees without edge lengths. So $W^0(n) = WA^0(n) = 1$, the tree with vertex label $m_n \in A(n)$. Map that one point into the map $[0,1] \longrightarrow [0,n]$ given by multiplication by n.

Now assume $f(n,k)\colon WA^k(n) \longrightarrow D(n)$ are given for $k \leq m$. $f(n, m+1)$ is defined on all trees with $m+1$ internal edges some one of which is 0 or 1 by relation a) and the operad morphism conditions. So we have $f(n, m+1)\Big|_{W^{m+1}(n) \times \partial I^{m+1}}$. This extends to all of I^{m+1} since $D(n)$ is contractable. So by induction, we get the operad morphism. Showing that the map commutes with μ is trivial, since that is built into the construction of the map on the boundary of I^{m+1}.

Remark. Comparing the three theories from an esthetic point of view, D has the advantage of being easiest to describe, and is useful for categories where lack of associativity is caused by combining homotopies. WA has the advantage of being more immediately generalizable to other theories. Finally, K has the honor (as far as I can tell) of being the first. The K(n) also have the advantage of being finite complexes.

Section 3. Categories and cotriples as monoids and their homotopy associate analogues.

In this section categories will be defined such that an A - object will be a category or a cotriple. A homotopy associative category or cotriple will then be a D - object in the appropriate category. Finally, it is shown that the usual relation between cotriples and categories -- the Kleisli construction -- generalizes to the homotopy associative case.

Definition 11. The category TGR(Q) is defined as follows for a fixed class Q: Its objects consist of a collection of topological spaces x(a,b) indexed by $Q \times Q$; a morphism is a collection of continuous maps $f(a,b): x(a,b) \longrightarrow y(a,b)$. Denote an object of TGR(Q) by $x = \{X(a,b) \mid (a,b) \in Q \times Q\}$.

Definition 12. A bifunctor $\otimes$: $TGR(Q) \times TGR(Q) \longrightarrow TGR(Q)$ is defined on objects by

$$X \otimes Y(z,b) = \coprod_{z \in Q} X(a,z) \times Y(z,b),$$

and on maps similarly: $f \otimes g(a,b) = \coprod_{z \in Q} f(a,z) \times g(z,b).$

Remark. Allow topological spaces to have underlying classes instead of underlying sets to avoid any logical difficulties.

Proposition 13. $\otimes$ as defined in Definition 12 is an associative bifunctor with identity.

Proof. Clearly $\otimes$ is associative. The identity is given by the object 1 defined by $1(a,b) \quad Q \times Q$.

Proposition 14. An A - object in $(TGR(Q), \otimes)$ is a topological category.

Proof. A map from $1 \to X$ picks out identity maps, and the map from $X \times X \longrightarrow X$ gives a composition. Associativity follows from Definition 1, i as in Proposition 4.

Definition 15. A homotopy associative category is a D - object in $TGR(Q)$ for some Q.

Now let $\underline{X}$ be a topological category. Let $E(\underline{X}) = \{F:\underline{X} \to \underline{Y} \mid F$ is a cont. functor$\}$. $E(\underline{X})$ can be made into a category by using continuous natural transformations as morphisms.

Definition 16. An associative bifunctor is defined on $E(\underline{X})$ as follows: $F \otimes G = F \cdot G$, the composition. For $\alpha:F \to F'$, $\beta:G \to G'$ natural transformations,

$$(\alpha \otimes \beta)_X = \alpha_{G'(x)} \cdot F(\beta_X).$$

Remarks.

1. The topology on $E(\underline{X})(F,G) = \{\alpha:F \to G \mid \alpha$ a cont. nat. trans.$\}$ is the subset topology from $\coprod_{X \varepsilon |X|}$, $X(F(X),G(X))$, a natural transformation being considered as an element in the product.

2. $\otimes$ defines an associative bifunctor on $E(\underline{X})^{op}$ as well. It will also be denoted $\otimes$, but is given by the composition $\omega. x. (\underline{\omega} x \underline{\omega})$, where ω is the contravariant functor $E(\underline{X}) \longrightarrow E(\underline{X})^{op}$, and $\underline{\omega}:E(\underline{X})^{op} \longrightarrow E(\underline{X})$. The identity in both cases is the identity functor, which shall be denoted by 1.

Proposition 17. An A - object in $E(\underline{X})$ is a triple. An A - object in $E(\underline{X})^{op}$ is a cotriple. In both cases the associative bifunctor is $\otimes$.

Proof. The first part is identical to Proposition 14, and the second follows by duality.

Definition 18. A homotopy associative cotriple on $\underline{X}$ is a D- object in $E(\underline{X})^{op}$.

Now if $G:\underline{X} \to \underline{X}$ is a cotriple, there is a standard category associated with G, called the Kleisli category of G and denoted $\underline{X}_G$. For an arbitrary functor $H:\underline{X} \to \underline{X}$, define an object of $TGR(|\underline{X}|)$ by $KH(X,Y) = \underline{X}(HX,Y)$. This will be called the Kleisli graph of H. [Kleisli] proves that the comultiplication of G, for G a cotriple, defines a composition in KG. This is summarized as:

Proposition 19. If G is a cotriple, KG is a category, with the composition of two maps $f:GX \to Y$, $g:GY \to Z$ given by $GX \xrightarrow[\delta_X]{} G^2X \xrightarrow[Gf]{} GY \xrightarrow[g]{} Z$.

Proposition 20. If G is a homotopy associative cotriple, KG is a homotopy associative category.

Remarks. It will be convenient in the following proofs to have an explicit description of the μ in End (G)(n) and End (KG)(n).

1. End (G). First denote $\alpha \otimes \beta$ by $\alpha G' \cdot G\beta$. By naturality that is the same as $\beta G \cdot F'\alpha$. So now suppose

$\alpha_i : G \longrightarrow G^i$, $i = 1,\cdots,k$; $\beta : G \longrightarrow G^k$. Then,

$$\mu(\alpha_1,\alpha_2,\cdots,\alpha_k,\beta) = \alpha_1 \otimes \alpha_2 \otimes \cdots \alpha_{k-1} \otimes \alpha_k \cdot \beta$$

$$= \alpha_1 G^{\sum\limits_{j=2^n}^{k} j} \cdot G\alpha_2 G^{\sum\limits_{j=3^n}^{k} j} \cdots G^{i-1}\alpha_i G^{\sum\limits_{j=(i+1)^n}^{k} j} \cdots G^{k-1}\alpha_k \cdot \beta .$$

2. End (KG). Let $f_i : KG^{(\dim f_i)} \longrightarrow KG$, $f : KG^{(k)} \longrightarrow KG$,

$i = 1,\cdots,k$. Then

$$\mu(f_1,\cdots,f_k,g)(a,b) : \coprod_{x_j} KG(a,x_1)\cdots x\ KG(x_{\Sigma n_i},b) \longrightarrow KG(a,b)$$

is given by

$$(f_1,\cdots,f_k,g)(a,b) = g\Big(f_1\big(h_1,h_2,\cdots,h_{n_1}\big), f_2\big(h_{n_1+1},\cdots,h_{n_1+n_2}\big),$$

$$\cdots, f_i\Big(h_{\left(\sum\limits_{j=1^n}^{i-1} j\right)+1},\cdots,h_{\sum\limits_{j=1^n}^{i} j}\Big),\cdots,f_k\big(,\cdots,h_{\Sigma n_i}\big)\Big).$$

<u>Proof</u>. Since G is a homotopy associative cotriple, there are continuous maps:

$$a(n) : D(n) \longrightarrow \{\alpha : G \longrightarrow G^n | \alpha \text{ is a cont. nat. trans.}\}, = \text{End } (G)(n)$$

which commute in the appropriate fashion. Define

$$b(n) : \text{End } (g)(n) \longrightarrow \text{End } (KG)(n)$$

by

$$b(n)(\alpha) : \coprod KG(X,X_1)_{-1}\times KG(X_1,X_2),\cdots,\times KG(X_{n-1},X_n) \longrightarrow KG(X,X_n)$$

$$b(n)(\alpha)(f_1,\cdots,f_n) = f_n \cdot Gf_{n-1} \cdot G^2f_{n-2},\cdots,G^{n-1}f_1 \cdot X$$

where $f_i : GX_{i-1} \longrightarrow X_i$, $i = 1,\cdots,nX_0 = X$.

$a(n)$ is already given as a morphism of operads. So if $b(n)$ is a morphism of operads, then $c(n) = b(n) \circ a(n)$ will also be such a morphism, and give KG the required structure.

To get $b(n)$ a morphism, it must be shown to commute with μ, i.e., suppressing indexes:

(1) $$\mu(bf_1,bf_2,\cdots,bf_k,bg) = b(\mu(f_1,f_2,\cdots,f_k,g)).$$

So let $\alpha_1,\cdots,\alpha_k,\beta$ be natural transformation of the appropriate dimensions. Let $\dim(\alpha_i) = n_i$, $\dim(\beta) = k$. Then

$$\mu(\alpha_1,\alpha_2,\cdots,\alpha_k,\beta) = \alpha_1 \otimes \alpha_2 \otimes \cdots \otimes \alpha_k \cdot \beta$$

$$= \gamma$$

$$= \alpha_1 G^{\sum_{j=2}^{k} n_j} \cdot G\alpha_2 G^{\sum_{j=3}^{k} n_j} \cdots G_{i-1}\alpha_i G^{\sum_{j=i+1}^{k} n_j} \cdots G^{k-1}\alpha_k \cdot \beta.$$

So $b(\Sigma n_i)(\mu(\alpha_1,\cdots,\alpha_k,\beta))(f_1,\cdots,f_m)$, where $m = \Sigma n_i$, is given by

$$f_m \cdot Gf_{m-1}, \cdots, G^{m-1} f_1 \cdot Y_X \, .$$

So the right hand side of (1) can be written as:

$$f_m \quad Gf_{m-1}, \cdots, G^{m-1} f_1 \cdot \alpha_1 G^{p_2} \quad G \, \alpha_2 \, G^{p_3}, \cdots, G^{k-1} \alpha_k \cdot \beta X$$

(2)

$$= f_m, \cdots, G^{m-1} f_1 \cdot \alpha_{1_{G^{p_2}(X)}} \left(G \, \alpha_{2_{G^{p_3}(X)}} \right) \cdots \; G^{k-1}(\alpha \, k_X) \cdot \beta_X$$

where

$$p_i = \sum_{j=1}^{k} n_j \, .$$

Now by naturality $G^{m-1} f_1 \cdot \alpha_{1_{G^{p_2}(X)}} = \alpha_{1_{G^{p_2 - 1}(X)}} \cdot G^{p_2 - 1}(f_1)$.

Now the left hand side of (1) is

$$b(\beta)(b(\alpha_1)(f_1, \cdots, f_{n_1}), \cdots, b(\alpha_i)\left(f_{\sum_{j=1}^{i-1} n_j + 1}, \cdots, f_{\sum_{j=1}^{i} n_j}\right),$$

$$\cdots, b(\alpha_k)\left(f_{m-n_k+1}, \cdots, f_m\right)$$

$$= b(\alpha_k)(f \cdots f) \cdot G(b(\alpha_{k-1})(f \cdots f)), \cdots, G^{k-1}(b(\alpha_1)(f \cdots f)) \cdot \beta_X$$

(3)

$$= f_m \cdot G(f_{m-1}), \cdots, G^{n_k - 1} f_{m-n_k - 1} \cdot \alpha_{k_X}, \cdots,$$

$$G^{i-1}(\alpha_{k-i+1_X}) \cdot G^i f\left(_{\sum_{j=1}^{k-i} n_m} \cdot Gf_{\sum_{j=1}^{k-i} n_j - 1}, \cdots, G^{k-1} f_{\sum_{j=1}^{k-i-1} n_j + 1}\right.$$

$$\left. \cdot \alpha_{k-1_X}\right), \cdots, X \cdot$$

We leave to the reader the straightforward but tedious verification that repeated application of the naturality rules gives a transformation of the expression (2) and (3).

Section 4. Examples

Example 1: Fibre spaces with transport ([Stasheff 2], [Malraison])

Think of a fibre space as a projection $p:E \to B$ with a path lifting function $\lambda:E \times_B MB \longrightarrow ME$, where $M(\ )$ is the space of Moore paths. An arbitrary fibre preserving map need not preserve the lifting, but it does do so up to homotopy. i.e., if $f:E \to E'$ is fibre preserving, the diagram

$$\begin{array}{ccc} E \times_B MB & \xrightarrow{\lambda} & ME \\ f \times id \downarrow & & \downarrow M f \\ E' \times_{B'} MB' & \xrightarrow[\lambda']{} & ME' \end{array}$$

commutes up to homotopy.

So take as maps fibre preserving maps and homotopies of the above diagram. Composition is obvious and clearly homotopy associative. Passing to the category with the same objects and homotopy classes of maps, one obtains the usual category of fibrations and fibre preserving maps. Two fibrations which have the same projection but different liftings will be different but isomorphic in the homotopy category.

Restricting to one base space B, one has a triple. Use of Lada's s.h.a. algebras gives another homotopy associative category, and loosens some of the restrictions on the lifting functions, e.g., they need not be regular.

Example 2: Homotopy invariant structures

If $\underline{B}$ is a topological theory, $W\underline{B}$ is the theory of spaces with the homotopy type of B - spaces. Every WB - space imbeds in a B - space as a deformation retract. [Boardman] If WB - spaces are made into a category by using maps which preserve the $\underline{WB}$ structure strictly, this procedure gives a functor $U : \underline{Top}^{WB} \longrightarrow \underline{Top}^{B}$.

If U is restricted to $\underline{Top}^{B}$, it has the structure of a homotopy associative cotriple. The corresponding Kleisli category provides one possible category of B - spaces and homotopy homomorphisms.

Example 3: Homotopy simplicial objects

Two special cases first: writing out the obcious (?) higher homotopies for a map between homotopy associative H - spaces, one encounters not a pentagon, but a hexagon:

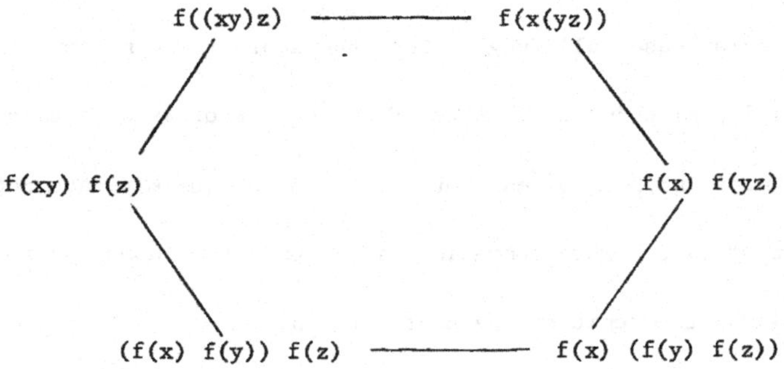

Secondly, in writing out higher homotopies for a homotopy T - space, for T a topological triple, one first requires that $\beta : \mu \simeq \beta . T\beta$. At the next stage, one encounters again a hexagon:

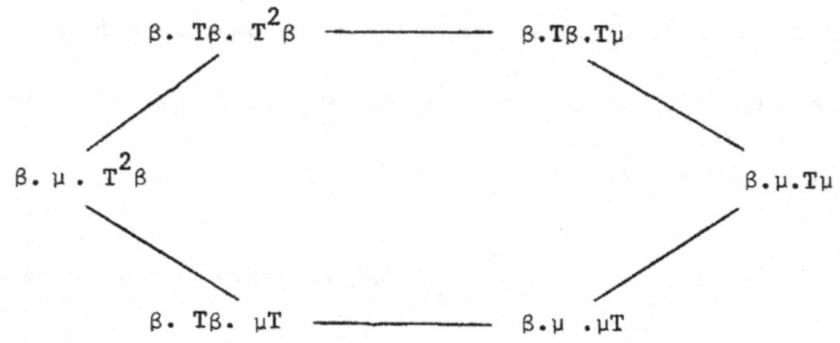

In the second special case above, β is the structure map of the T - space and μ is the multiplication of the triple. One of the sides of the hexagon is not a homotopy but an equality, because of naturality of μ . A similar diagram arises when considering what a homotopy associative triple ought to be directly (i.e., starting with the single homotopy associativity and building up higher ones).

These special cases all suggest that the appropriate notion is a homotopy simplicial object. This could be thought of as a sequence of spaces with maps and homotopies, but such a notion quickly gets notationally out of hand. More convenient is a concept of homotopy functor, and that will be the topic of the next section.

Section 5. Homotopy Associative Functors

In trying to discover in what sense a homotopy associative triple arises from a pair of "adjoint" functors, and in trying to get a convenient notion of homotopy simplicial object, one is confronted with the need for a concept of functor which commutes with the structures up to homotopy, i.e., some sort of strongly homotopy multiplicative functor. Abusing language, call such a concept a homotopy associative functor.

The most immediate definition would be a D-object in the category $TGR^{\underline{2}}$ of maps between topological graphs. A few examples show that that is too strong, i.e., requiring strict commutativity in some cases where only commutativity up to homotopy is given.

The next possible approach is to build up the notion inductively: a homotopy associative functor is a map on objects, a map on maps, a homotopy for each pair of maps between $F(fg)$ and $F(f)F(g)$, etc. One actually, needs a homotopy for each pair of maps, and each element of $D(2)$ and similarly for the higher homotopies. The result obtained is similar to Lada's s.h. algebra maps in the case where either the domain or range is a category.

A third possibility is to define simplicial objects up to homotopy as collections of spaces and homotopies, and maps between such things similarly, and then re-interpret categories and functors as simplicial objects.

All of these approaches have advantages and disadvantages, and rather than going into the details, I would like to conclude by listing some properties one might like homotopy associative functors to satisfy.

1. The collection of homotopy associative functors is a homotopy associative category. This necessitates the inevitable messy question of homotopy natural transformations. The first construction given above (D-objects in the category of maps) seems to satisfy this condition.

2. Homotopy simplicial objects should be realizable as is. One could also pass to the homotopy category, get a strict simplicial object and realize in the homotopy category, if possible. If everything up to this point works, does realizing and passing to the homotopy category yield the same result as passing to homotopy and realizing?

3. Homotopy associative (co)triples arise from adjoint pairs of functors.

4. Homotopy associative functors should be the right way to formulate homotopy limits [Bousefield-Kan]. Is there a homotopy analogue of the adjoint functor theorem?

Added in proof: Saunders MacLane points out that non-assciative categories were studied by Benabou under the name bicategories, in Volume I of the Reports of the Midwest Category Seminar.

References

Boardman, J. M., Homotopy Structures and the Language of Trees, AMS Proc. Symp. Pure Math., Vol. XXII, p. 37-58.

Boardman, J. M. and Vogt, R., Homotopy Everything H-spaces, Bull. Amer. Math. Soc. 74 (1968), p. 1117-1122.

Bousfield, A. and Kan, D., "Homotopy Limits, Completion and Localization", Lecture Notes in Mathematics, Vol. 304, Springer, Berlin, 1973.

Kleisli, H., Every Standard Construction is an Adjoint Pair, Proc. Amer. Math. Soc. 16 (1965), p. 544-46.

MacLane S., Categorical Algebra, Bull. Amer. Math. Soc. 71 (1965), p. 40-106.

Malraison, P., Fibrations as Triple Algebras, Journal of Pure and Applied Algebra, to appear.

May, J. P., "The Geometry of Iterated Loop Spaces," Lecture Notes in Mathematics, Vol. 271, Springer, Berlin, 1972.

Stasheff J., 1. Homotopy Associativity of H-spaces, I, II, Trans. Amer. Math. Soc. 108 (1963), p. 293-312.

2. Parallel transport in Fibre Spaces, Bol. Soc. Math. Mex. 1966, p. 68-84.

Lada, T., Strong Homotopy D-algebras, AMS Notices, Vol. 20, No. 1, Abstract 701-55-14 and preprint/thesis.

Characteristic classes
and K-theory

J. C. Becker*

1. Introduction. If $p\colon E \to B$ is a fiber bundle with fiber F, there is an S-map $\tau\colon B^+ \to E^+$ having the fundamental property that the composite

$$H^*(B) \xrightarrow{p^*} H^*(E) \xrightarrow{\tau^*} H^*(B),$$

in singular cohomology, is multiplication by $\chi(F)$ - the Euler characteristic of F [3]. In this talk we will describe some applications of this map, which we call the transfer, to K-theory.

Let O denote the infinite orthogonal group over either the real, complex or quaternionic numbers. We will use k^* to denote the cohomology theory obtained from the Bott spectrum by replacing the n-th space of the spectrum by its n-connected cover. Thus, $k^O(X) = [X;BO]$. Let $\varepsilon = 1$ in the complex and symplectic case and $\varepsilon = 2$ in the real case. Let $\lambda\colon BO(\varepsilon) \to BO$ denote the inclusion. It determines a transformation of cohomology theories on the category of pointed CW-complexes

$$\lambda_*\colon \{\quad ;BO(\varepsilon)\}^* \longrightarrow k^*,$$

where $\{X,Y\}$ denotes the group of stable maps from X to Y [15].

THEOREM 1. $\lambda_*\colon \{\quad ; BO(\varepsilon)\} \longrightarrow k^0$ is epimorphic.

The complex case of this theorem is due to G. Segal [13].

Suppose now that h^* is a cohomology theory and $c\colon k^O \to h^O$ is a characteristic class such that (a) c extends to a trans-

* This talk is based on joint work with D. Gottlieb.

formation of cohomology theories and (b) the composite

$$[\quad ; BO(\epsilon)] \xrightarrow{\lambda} k^{o} \xrightarrow{c} h^{o}$$

is zero. Then because of (a) we actually have that

$$\{\quad ; BO(\epsilon)\} \xrightarrow{\lambda} k^{o} \xrightarrow{c} h^{o}$$

is zero, whence by theorem 1, c is identically zero.

It is well known that the Adams conjecture can be viewed as a statement involving a transformation of cohomology theories. Precisely, let BF be the classifying space for spherical fibrations and let $Sph(X) = [X; BF]$.

COROLLARY 1. [7], [11], [14], [3]. The composite

$$k_R^o(X) \longrightarrow k_R^o(X) \otimes Z[t^{-1}] \xrightarrow{\psi^t - 1} k_R^o(X) \otimes Z[t^{-1}] \xrightarrow{J} Sph(X) \otimes Z[t^{-1}],$$

is trivial for every finite complex X.

Letting c denote this composite we must check that c satisfies conditions (a) and (b). Because of the relation between ψ^t and the Bott map [1], ψ^t extends to a transformation of cohomology theories.

$$\psi^t\colon k_R^*(\quad) \otimes Z[t^{-1}] \longrightarrow k_R^*(\quad) \otimes Z[t^{-1}] .$$

Now Boardman and Vogt [4] have shown that BO and BF are infinite loop spaces and $J\colon BO \longrightarrow BF$ is an infinite loop map. Moreover, May [10] has shown that this infinite loop structure on BO is equivalent to the one obtained from the Bott spectrum, i.e. the spectrum which defines k^*. Thus, J also extends to a transformation of cohomology theories. Hence c satisfies condition (a). Finally, Adams [1] has shown that c satisfies condition (b).

Since $\{\quad ; BO(\epsilon)\} = [\quad ; Q(BO(\epsilon))]$ where $Q(BO(\epsilon)) = \Omega^\infty S^\infty(BO(\epsilon))$, theorem 1 may be rephrased to state that the natural extension $\bar{\lambda}\colon Q(BO(\epsilon)) \longrightarrow BO$ of λ induces an epimorphism.

$$\bar{\lambda}_{\#}: [X; Q(BO(\varepsilon))] \longrightarrow [X; BO]$$

for every CW-complex X. Letting $X = BO$ we obtain a right inverse for $\bar{\lambda}$. Now, since $\bar{\lambda}$ is an H-map it is easy to see that $Q(BO(\varepsilon))$ is homotopy equivalent to $BO \times F$ where F is the fiber of $\bar{\lambda}$. Comparing the rank of $\Sigma_*(BO(\varepsilon))$ with that of $\pi_{\#}(BO)$ we see that F has finite homotopy groups. Hence we have

COROLLARY 2. *There are direct product decompositions*

$$Q(BO(2)) \simeq BO \times F$$
$$Q(BU(1)) \simeq BU \times F'$$
$$Q(BSP(1)) \simeq BSP \times F''$$

where F, F' *and* F'' *have finite homotopy groups.*

2. *The transfer.* By a fiber bundle $\xi = (E,B,p)$ we mean one whose base B is a finite complex, whose structure group G is a compact Lie group and whose fiber F is a compact smooth G-manifold without boundary. If B is also a smooth manifold without boundary the transfer

$$\tau(\xi): B^+ \longrightarrow E^+$$

is constructed as follows. Let $E \subset B \times S^s$ be an embedding homotopic to p and having normal bundle β. Then β is inverse to the bundle of tangents along the fiber α and the embedding yields a trivialization $\psi: \alpha \oplus \beta \longrightarrow E \times R^s$. The transfer is then represented by the composite

$$B^+ \wedge S^s \xrightarrow{\ t\ } E^\beta \xrightarrow{\ i\ } E^{\alpha\oplus\beta} \xrightarrow{\ \psi\ } E^+ \wedge S^s ,$$

where t is the Pontryagin-Thom map and i is the inclusion (see [3]). If ξ is a finite covering, $\tau(\xi)$ is the same as the transfer of Kahn and Priddy [9] and Roush [12].

The transfer is functorial with respect to bundle maps and obeys the product formula

$$\tau^*(p^*(x)y) = x\tau^*(y) \quad , \quad x \in H^*(B), \quad y \in H^*(E).$$

In order to show that

$$H^*(B) \xrightarrow{p^*} H^*(E) \xrightarrow{\tau^*} H^*(B)$$

is multiplication by $\chi(F)$, it is sufficient, in view of the above properties to consider the trivial bundle $F \to pt$. Thus, if $F \subset R^s$ with normal bundle β we must show that

$$S^s \xrightarrow{t} F^\beta \xrightarrow{i} F^{\alpha\oplus\beta} \xrightarrow{\psi} F^+ \wedge S^s \xrightarrow{\text{proj.}} S^s$$

has degree $\chi(F)$. As pointed out in [3] this is essentially the well known Hopf vector field theorem.

We shall also need the following property of the transfer. Let F be a G-manifold and $F' \subset F$ a G-invariant submanifold. Let W be a closed G-invariant tubular neighborhood of F' [6] and assume that $\overline{F - W}$ has a non zero G-equivariant vector field Δ whose restriction to $\dot{W}$ lies in the tangent space of $\dot{W}$. If ξ is a fiber bundle with fiber F let ξ' denote the subbundle with fiber F', and $i: \xi' \longrightarrow \xi$, the inclusion.

LEMMA 1. <u>With the above assumptions</u>

$$\begin{array}{ccc} & \xrightarrow{\tau(\xi)} & E^+ \\ B^+ & & \uparrow i \\ & \xrightarrow{\tau(\xi')} & (E')^+ \end{array}$$

<u>is commutative</u>.

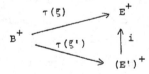

In particular, suppose that S^1 acts on F and the action commutes with G, so that we have an action of $S^1 \times G$ on F. Let F' denote the fixed point set of the S^1-action. Then the hypotheses of the lemma are satisfied for we may take W to be a closed $S^1 \times G$ - invariant tubular neighborhood of F' and Δ the vector field on $\overline{F - W}$ determined by the S^1-action.

3. <u>Coset spaces</u>. Let T be a maximal torus of a connected compact Lie group G and N the normalizer of T in G. By a theorem of Hopf and Samelson [8] [5] the Euler characteristic of G/N is 1.

Let $\mathcal{S}_n$ denote the symmetric group and consider the wreath product subgroups

$$\mathcal{S}_n \int O(2) \subset O(2n)$$

$$\mathcal{S}_n \int U(1) \subset U(n)$$

$$\mathcal{S}_n \int Sp(1) \subset Sp(n)$$

LEMMA 2. The coset spaces $O(2n)/\mathcal{S}_u \int O(2)$, $U(n)/\mathcal{S}_n \int U(1)$ and $Sp(n)/\mathcal{S}_n \int Sp(1)$ have Euler characteristic 1.

<u>Proof</u>. These all follow from the Hopf-Samelson theorem. In the real case $O(2n)/\mathcal{S}_n \int O(2) = SO(2u)/N_o$ where N_o is the normalizer of the standard torus in $SO(2n)$. In the complex case $\mathcal{S}_n \int U(1)$ is the normalizer of the standard torus in $U(n)$. In the quaternionic case, the normalizer of the standard torus is $\mathcal{S}_n \int Pin(2)$ where $Pin(2) \subset SP(1)$ is the normalizer of S^1 in $SP(1)$. Recall that if $F \to E \to B$ is a fiber bundle, $\chi(E) = \chi(F)\chi(B)$. Applying this to the bundle

$$\frac{\mathcal{S}_n \int SP(1)}{\mathcal{S}_n \int Pin(2)} \longrightarrow \frac{SP(n)}{\mathcal{S}_n \int Pin(2)} \longrightarrow \frac{SP(n)}{\mathcal{S}_n \int SP(1)} ,$$

we see that $\chi\left(SP(n)/\mathcal{S}_n \int SP(1)\right) = 1.$

From this point on the proof of theorem 1 is identical in the real, complex, and quaternionic cases. For simplicity of notation we restrict our attention to the real case.

As a specific model for BO we will take

$$BO = \mathrm{Lim}_n \; O(4n)/O(2n) \times O(2n) \quad ,$$

where the inclusion $O(4n) \subset O(4n+4)$ is compatible with the inclusion $R^{4n} \subset R^{4n+4}$ given by

$$(x_1,\ldots,x_{4n}) \longrightarrow (x_1,\ldots,x_{2n},0,0,x_{2n+1},\ldots,x_{4n},0,0) \quad .$$

Let

$$BN = \mathrm{Lim}_n \; O(4n)/\mathcal{S}_n \int O(2) \times O(2n)$$

and let $p\colon BN \to BO$ be the projection.

THEOREM 2. For any cohomology _theory_ h, $p^*\colon h^*(BO) \longrightarrow h^*(BN)$ _is a monomorphism_.

Proof. Let

$$B_n = O(4n)/O(2n) \times O(2n) \quad ,$$

$$E_n = O(4n)/\mathcal{S}_n \int O(2) \times O(2n) \quad ,$$

and $p_n\colon E_n \longrightarrow B_n$ the projection. It has a transfer $\tau_n\colon B^+_n \longrightarrow E^+_n$. We will show now that the square

$$(3.1) \qquad \begin{array}{ccc} E^+_n & \longrightarrow & E^+_{n+1} \\ \uparrow{\scriptstyle \tau_n} & & \uparrow{\scriptstyle \tau_{n+1}} \\ B^+_n & \longrightarrow & B^+_{n+1} \end{array}$$

is commutative.

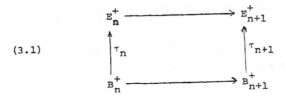

The pullback of E_{n+1} over B_n has fiber $F = O(2n+2)/\mathcal{S}_{n+1}\int O(2)$ and structure group $O(2n)$ acting by left translation. Identify S^1 with the subgroup

$$\{1\}\times\ldots\times\{1\} \times SO(2) \subset O(2n+2)$$

and let S^1 act on F by left translation. It commutes with the action of $O(2n)$. To determine the S^1 fixed point set let $t \in S^1$ be a generator, $y \in O(2n+2)$ and $[y] \in F$ the left coset containing y. If $t[y] = [y]$ then $ty = yn$, $n \in \mathcal{S}_{n+1}\int O(2)$. Hence by $y^{-1}S^1y \subset \mathcal{S}_{n+1}\int O(2)$. This implies that $[y] \in O(2n)/\mathcal{S}_n\int O(2)$. Thus, the S^1-fixed point set is $O(2n)/\mathcal{S}_n\int O(2)$. The commutativity of (3.1) now follows from lemma 1 and the remarks following.

We may now construct a stable map (i.e. a map in the stable homotopy category) $\tau\colon BO^+ \longrightarrow BN^+$ whose restriction to B_n^+ is homotopic to τ_n for all n. Since the Euler characteristic of $O(2n)/\mathcal{S}_n\int O(2)$ is 1,

$$\left(p_n^+ \tau_n\right)_* : \tilde{H}_*(B_n^+) \longrightarrow \tilde{H}_*(B_n^+)$$

is an isomorphism. Therefore

$$(p^+\tau)_* : \tilde{H}_*(BO^+) \longrightarrow \tilde{H}_*(BO^+)$$

is an isomorphism, and $p^+\tau : BO^+ \longrightarrow BO^+$ is a stable homotopy equivalence. Hence, if h is any cohomology theory

$$(p^+\tau)^* : h^*(BO^+) \longrightarrow h^*(BO^+)$$

is an isomorphism; whence $(p^+)^*$ is a monomorphism.

Since the horizontal maps in the diagram

$$\begin{array}{ccc}
h^*(BN) & \longrightarrow & h^*(BN^+) \\
\uparrow{\scriptstyle p^*} & & \uparrow{\scriptstyle (p^+)^*} \\
h^*(BO) & \longrightarrow & h^*(BO^+)
\end{array}$$

are monomorphisms, p^* is also a monomorphism.

4. Proof of theorem 1. The argument is essentially that of [3] expanded so as to apply to infinite complexes. We begin by showing that

$$(4.1) \qquad p_* : \{\quad ,BN\} \longrightarrow k^o \quad ,$$

induced by $p: BN \longrightarrow BO$, is epimorphic. Let h be the mapping cone theory of the transformation $p_* : \{\quad ,BN\}^* \longrightarrow k^*$. We have an exact sequence

$$\cdots \longrightarrow \{\quad ,BN\} \xrightarrow{p_*} k^o \xrightarrow{c} h^o \longrightarrow \cdots$$

Let $u \in k^o(BO)$ be the class of the identity map. It is sufficient to show that u is in the image of p_*, that is, that $c(u) = 0$. We have

$$\begin{array}{ccccc}
\{BO,BN\} & \xrightarrow{p_*} & k^o(BO) & \xrightarrow{c} & h^o(BO) \\
\downarrow{\scriptstyle p^*} & & \downarrow{\scriptstyle p^*} & & \downarrow{\scriptstyle p^*} \\
\{BN,BN\} & \xrightarrow{p_*} & k^o(BN) & \xrightarrow{c} & h^o(BN)
\end{array} .$$

Since $p^*(u) = p_*\big([1_{BN}]\big)$, $c(p^*(u)) = 0$. Since p^* is monomorphic we have $c(u) = 0$.

If $[q] \in \{BO,BN\}$ is such that $p_*([q]) = u$ then q determines a right inverse for p_*,

$$(4.2) \qquad q_* : k^o \longrightarrow \{\quad ,BN\}$$

LEMMA 3. <u>There is a stable map</u> $\zeta: BN \longrightarrow BO(2)$ <u>such that</u>

$$\begin{array}{ccc} & & \{X, BO(2)\} \\ & \overset{\zeta_*}{\nearrow} & \\ \{X;BN\} & & \downarrow \lambda_* \\ & \underset{p_*}{\searrow} & \\ & & k^o(X) \end{array}$$

<u>is commutative for any finite complex</u> X.

Assuming the lemma, consider the maps

$$k^o \xrightarrow{q_*} \{\quad, BN\} \xrightarrow{\zeta_*} \{\quad, BO(2)\} \xrightarrow{\lambda_*} k^o.$$

By the lemma, this composite is an isomorphism for any finite complex; hence for any complex. Therefore λ_* is epimorphic, proving theorem 1.

To prove lemma 3, represent $BO(2)$ as

$$BO(2) = \mathrm{Lim}_n\, O(4n)/O(2) \times O(2n-2) \times O(2n)$$

and let

$$\tilde{E}_n = O(4n)/O(2) \times \Sigma_{n-1}\int O(2) \times O(2n).$$

The quotient map $r_n: \tilde{E}_n \longrightarrow E_n$ is an n-fold covering. Let $\tau_n : E_n^+ \longrightarrow \tilde{E}_n^+$ denote its transfer and let

$$\zeta_n' : E_n^+ \longrightarrow BO(2)$$

be given by

$$E_n^+ \xrightarrow{\pi_n} \tilde{E}_n^+ \xrightarrow{\pi_n} O(4n)/O(2) \times O(2n-2) \times O(2n) \longrightarrow BO(2),$$

where π_n is the quotient map. It has the following properties

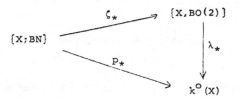

(4.3)
$$\begin{array}{ccc} E_n^+ & \xrightarrow{\zeta_n} & \\ \downarrow & & BO(2) \\ E_{n+1}^+ & \xrightarrow{\zeta'_{n+1}} & \end{array}$$

is commutative.

(4.4)
$$\begin{array}{ccc} \{X,E_n^+\} & \xrightarrow{(\zeta_n')_*} & \{X,BO(2)\} \\ & \searrow^{(p_n')_*} & \downarrow \lambda_* \\ & & k^o(X) \end{array}$$

<u>is commutative for any finite complex</u> X.

Here $p_n'\colon E_n^+ \longrightarrow BO$ is the extension of $p_n\colon E_n \longrightarrow BO$.

Property (4.3) is easily checked after observing that the inverse image of E_n under $r_{n+1} : \widetilde{E}_{n+1} \longrightarrow E_{n+1}$ consists of $\widetilde{E}_n$ plus a disjoint copy of E_n, and this disjoint copy is mapped to the base point by π_{n+1}.

To prove (4.4) it is sufficient to take $X = E_n^+$ and $u \in \{E_n^+,E_n^+\}$ the class of the identity. Now $k^o(E_n^+) = [E_n^+;BO]$ is the group of stable equivalence classes of vector bundles over E_n and $(p_n')_*(u)$ is the class (α) of the vector bundle α with fiber R^{2n} associated with the principal $\mathcal{S}_n\int O(2)$-bundle

$$O(4n)/\{1\} \times O(2n) \longrightarrow O(4n)/\mathcal{S}_n\int O(2) \times O(2n) = E_n .$$

Let β denote the vector bundle with fiber R^2 associated with the principal $O(2)$-bundle

$$O(4n)/\{1\} \times \mathcal{S}_{n-1}\int O(2) \times O(2n) \longrightarrow O(4n)/O(2) \times \mathcal{S}_{n-1}\int O(2) \times O(2n) = \widetilde{E}_n .$$

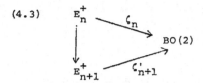

Then $\lambda_*(\zeta_n')_*(u)$ is the image of β under the transfer map

$$\tau_n^* : k^o(\tilde{E}_n^+) \longrightarrow k^o(E_n^+) .$$

Thus we must show that $\tau_n^*(\beta) = (\alpha)$. This can be done directly as in [3; section 7] using the geometric description of K-theory transfer as the "direct image" map.

Because of (4.3) we may construct a stable map

$$\zeta' : BN^+ \longrightarrow BO(2)$$

whose restriction to E_n^+ is homotopic to ζ_n' for each n. Let $\varepsilon : BN \longrightarrow BN^+$ be a stable map which is a (base point preserving) right inverse for the projection $BN^+ \longrightarrow BN$. Then $\zeta = \zeta'\varepsilon : BN \longrightarrow BO(2)$ satisfies the requirements of lemma 3.

REFERENCES

1. Adams, J. F., Vector fields on spheres, Ann. of Math. Ann. of Math. 45 (1962), 603-632.

2. ___________, On the groups J (X) - I, Topology 2 (1963, 181-195.

3. Becker, J. C. and Gottlieb, D. H., The transfer map and fiber bundles, Topology, to appear.

4. Boardman, J. M. and Vogt, R. M., Homotopy everything H-spaces, Bull. Amer. Math. Soc. 74 (1968). 117-1122.

5. Bredon, G. E., Introduction to Compact Transformation Groups, Academic Press, New York, 1972.

6. Conner, P. E. and Floyd, E.E., Differentiable Periodic Maps, Academic Press, New York, 1964.

7. Friedlander, E., Fibrations in etale homotopy theory, Publications Mathematiques I.H.E.S. 42 (1972), 1-46.

8. Hopf, H. and Samelson, H., Ein satz uber die winkungsraume geschlossener Lie'scher gruppen, Comm. Math. Helv. 13 (1940), 240-251.

9. Kahn, D. S. and Priddy, S. B., Applications of the transfer to stable homotopy theory, Bull. Amer. Math. Soc. 78 (1972), 981-987.

10. May, J. P., $\mathcal{J}$-functors and orientation theory, to appear.

11. Quillen, D., The Adams conjecture, Topology 10 (1971), 67-80.

12. Roush, F. W., Transfer in generalized cohomology theories, Thesis, Princeton University, 19711.

13. Segal, G., The stable homotopy of complex projective space, Quart. Jour. of Math. 24 (1973), 1-5.

14. Sullivan, D., Geometric Topology, Part I, Localization, Periodicity and Galois Symmetry, mimeographed M.I.T., 1970.

15. Vogt, R. M., Boardman's stable homotopy category, mimeographed, Aarhus University, 1970.

THE STRUCTURE OF MAPS FROM MANIFOLDS TO SPHERES

by

Louis M. Mahony

Introduction. As is well known, if $f: M^{n+k} \longrightarrow S^n$ is a map from a compact smooth or piecewise linear manifold M^{n+k} to the n - sphere, it may be interpreted as a submanifold N^k with a particular trivilization F^n of its normal microbundle in M^{n+k}. (In the topological category, recent results of Kirby and Siebenmann show such an interpretation is also possible provided suitable dimension assumptions are made.) Under this interpretation, the homotopy class of f is translated into the framed cobordism class of the pair (N^k, F^n). With this description of a homotopy class, the problems investigated in this paper were: 1) Given a framed cobordism class, find a representative in this class which is localized in a particular nice region of the manifold M^{n+k}. For example, is it possible to find a framed manifold in the equivalence class which is contained in a disk? 2) To determine if a given class has a representative which is of a simple form such as a framed homotopy sphere.

Section 1. *Localizing Maps*. In this section, certain results are proved which enable one to localize the homotopy classes of maps of a topological manifold to a particular region of the manifold.

Definition 1.0. *Let* M^n *be a topological manifold.* M^n *is said to have a Smale-Wallace decomposition if the following conditions are satisfied.* (1) *There exists a sequence of submanifolds* $C^0 \subset C^1 \subset \cdots \subset C^{n-1} \subset C^n$ *with* C^0 *equal to* D^n, *the* n-*dimensional disk, and* C^n *equal to* M^n. (2) C^k *is obtained from* C^{k-1} *by attaching handles of index* k *only.* M^n *is obtained from* C^{n-1} *by attaching a single* n *cell* D^n.

In the rest of this paper, once a decomposition is chosen, it will be assumed fixed unless otherwise specified.

It is known [7], that if M^n admits a smooth or piecewise linear structure, such a Smale-Wallace decomposition is always possible. If $n \geq 5$, and M^n is a topological manifold, results of Kirby and Siebenman show that M^n has handle body decomposition.

It follows from the above, that there exists a dual decomposition denoted by $C_0 \supset C_1 \supset \cdots \supset C_{n-1} \supset C_n$ with $C_0 = M^n$, $C_n = D^n$, $C^k \cup C_{n-k-1} = M^n$, and $bC^k = bC_{n-k-1}$ where b stands for the boundary operator. Relative to the above Smale - Wallace decomposition, one can show that the homotopy classes of maps of M^n into S^p do not depend on certain higher dimensional handles. More precisely, one has the following.

Proposition 1.1. For $p > n-k$, there is a bijection q^* from $[M^n, S^p]$ to $[C^k/bC^k, S^p]$. Moreover, for $2p - 2 \geq n$, the cohomotopy group $[M^n, S^p]$ is isomorphic to $[C^{[n/2]}/bC^{[n/2]}, S^p]$.

The condition $2p-2 \geq n$ implies that $[M^n, S^p]$ has a natural abelian group structure, and the latter part of the above proposition says that the group structure depends on only half the handles. Also, X/A means the space obtained from X by collapsing $A \subset X$ to a point and $[n/2]$ denotes the integral part of $n/2$.

Proof. Consider the following cofibration,

$$C_{n-k-1} \xrightarrow{i} M^n \xrightarrow{q} M^n/C_{n-k-1}$$

where i, q are the inclusion and projection maps respectively. It is easily verified that M^n/C_{n-k-1} is homeomorphic to C^k/bC^k. From the above cofibration, one gets the Puppe sequence:

$$C_{n-k-1} \xrightarrow{i} M^n \xrightarrow{q} C^k/bC^k \xrightarrow{t} S(C_{n-k-1}) \xrightarrow{Si} S(M^n) \quad \cdots .$$

S denotes the suspension map. Applying the half exact functor $[\ ,S^p]$ to this, one obtains an exact sequence,

$$[C_{n-k-1}, S^p] \xleftarrow{i^*} [M^n, S^p] \xleftarrow{q^*} [C^k/bC^k, S^p] \xleftarrow{t^*} [S(C_{n-k-1}), S^p] \quad \cdots .$$

Since C_{n-k-1} has the homotopy type of a $n-k-1$ dimensional complex, $p > n-k$ implies that the two sets at the ends of the previous sequence are trivial. Since the sequence is exact, the first part of the proposition follows.

Under the hypothesis, $2p-2 \geq n$, if one takes for k in the above exact sequence, $k = [n/2]$ then one can show that again both groups at the ends are zero, and therefore, $[C^{[n/2]}/bC^{[n/2]}, S^p]$ is isomorphic to $[M^n, S^p]$.

If M^n is a smooth manifold, then the homotopy classes of maps from M^n to spheres can be interpreted, via the Thom-Pontriagin construction, as framed cobordism classes of framed submanifolds of M^n. An obstruction arises when one tries to move a framed submanifold into lower dimensional handles in the following way.

If the framed submanifold (N^{n-p}, F^p) represents a homotopy class, then N^{n-p} may have a non-trivial point intersection with the belt disk of the $n-p$ handles. This is the obstruction to pushing (N^{n-p}, F^p) down in the $n-p-1$ dimensional handles.

The following corollary shows this obstruction need not always exist.

Corollary 1.2. *Let M^{2n}, $n \geq 3$, be an $n-1$ connected $2n$ dimensional piecewise linear manifold. Then for $p > n+1$, the cohomotopy group $[M^{2n}, S^p]$ is isomorphic to $[S^{2n}, S^p]$. Moreover, for $p = n+1$, there is a homomorphism q^* of $[S^{2n}, S^p]$ onto $[M^{2n}, S^p]$.*

The corollary shows that if one takes a disk D^{2n} and an embedding

j of D^{2n} into M^{2n}, then this induces a monomorphism j^* of $[M^{2n}, S^p]$ onto $[S^{2n}, S^p]$ in the stated range.

Proof. From 1.1, the following exact sequence exists:

$$[C_n, S^p] \xleftarrow{i^*} [M^{2n}, S^p] \xleftarrow{q^*} [C^{n-1}/bC^{n-1}, S^p] \xleftarrow{t^*} [S(C_n), S^p] \quad \cdots .$$

It is shown in [7], that under the stated hypothesis, one may assume M^{2n} has only handles of index n. From this one can easily show that C^{n-1}/bC^{n-1} is homeomorphic to S^{2n}. Then for $p > n+1$ the two end groups in the above exact sequence are zero, and for $p = n+1$ only the left hand end group is zero, whence the assertions of the corollary.

Actually later on, the group $[M^{2n}, S^{n+1}]$ is determined if M^{2n} is smooth. See Corollary 1.9 .

Suppose now, M^{2k+j} is a smooth manifold with Smale - Wallace function f which has only one critical point of index 0, only one of index n and the only other critical points are of index k and $k+j$, $k \geq 2$, $j \geq 1$. Because C^k contains only critical points not larger than k, C^k is built up in the following way. There exist diffeomorphisms f_i from $S_i^{k-1} \times D_i^{k+j}$ into $S^{2k+j-1} = bD^{2k+j}$, $i = 1,\cdots s$, s = number of critical points of index k, with the f_i having disjoint images. C^k is the space obtained by identifying the subset $S_i^{k-1} \times D_i^{j+j}$ of $D_i^k \times D_i^{k+J}$ with its image in D^{2k+j}. Let $S^{k-1} \times D^{k+j}$ be the standard subset of $S^{2k+j-1} = b(D^k \times D^{k+j})$. Then there is an ambient isotopy of S^{2k+j-1} such that the image of $S_i^{k-1} \times D_i^{k+j}$, by the map f_i and the isotopy, is equal to $S^{k-1} \times D^{k+j}$. Such an isotopy determines for each map f_i, a map of S^{k-1} into SO_{k+j}. Since $j \geq 1$, one may identify, by Bott periodicity, $\pi_{k-1}(SO_{k+j})$ with the integers Z, Z_2, or 0. Hence, each such f_i determines a number m_i belonging to Z or Z_2 depending upon if $k \equiv 0 \mod (4)$ or $k \not\equiv 0 \mod (4)$. Let $(m_1, m_2, \cdots, m_s)$ denote this sequence associated to the f_i's. If J_{k-1} is the stable Hopf-Whitehead homomorphism from $\pi_{k-1}(SO)$ to π_{k-1}^S (the stable cohomotopy group of spheres), set $a(m_i) = J_{k-1}(m_i)$, and let $(a(m_1),\cdots,a(m_s))$ denote this sequence.

Corollary 1.3. *Let* M^{2k+j} *be a smooth manifold* $k \geq 2$, $j \geq 1$ *which admits a Smale - Wallace function with one maximum and one mimimum, and the only other critical points are of index* k *and* $k+j$. *Then for* $p > k+1$, $[M^{2k+j}, S^p]$ *has a natural abelian group structure. If* $k = 2, 4,$ or 8, *assume* $j \geq 2$.

The condition that $2p - 2 \geq 2k + j$ is not required in the above statement.

Proof of Corollary. From Proposition 1.1, it follows that for $p > k+1$, $[M^{2k+j}, S^p]$ is naturally isomorphic to $[C^k/bC^k, S^p]$. By using the technical Lemma 1.4, there exists a smooth manifold N^{2k+j-2} with boundary bN^{2k+j-2} such that $S^2(N^{2k+j-2}/bN^{2k+j-2})$ is homeomorphic to C^k/bC^k, whence the assertion.

Technical Lemma 1.4. *Given* C^k *as above, then there exists a smooth manifold* N^{k+p} *with boundary* bN^{k+p} *such that* C^k/bC^k is *homeomorphic to* $S^{k+j-p}(N^{k+p}/bN^{k+p})$ *where* S^{k+j-p} *is the* $k+j-p$ *interated suspension and*

(i) $p = [k/2] + 2$ _if_ $[k/2] \geq 13$

(ii) $p = k - 1$ _if_ $k = 2, 4, 8$

(iii) $p = k$ _otherwise_.

Proof. Let $(m_1, m_2, \cdots, m_s)$ be the sequence associated with M^{2k+j}, where $m_i \in \pi_{k-1}(SO_{k+j})$. In case (i), a statement of Barratt and Mahowald [2] asserts the existence of elements $\underline{m}_i \in \pi_{k-1}(SO_p)$ such that $S^{k+j-p}(\underline{m}_i) = m_i$ where S^{k+j-p} is the interated suspension of $S: SO_p \longrightarrow SO_{p+1}$. In case (ii), the existence of such elements $\underline{m}_i$ is given in Kosinski [11] and in case (iii), the $\underline{m}_i$ exists from the fact $S_*: \pi_{k-1}(SO_k) \longrightarrow \pi_{k-1}(SO_{k+1})$ is onto.

Let $C(k,p)$ be the manifold obtained by attaching s handles of index k and codimension p to the $k+p$ dimensional disk D^{k+p}, using maps $s_i: S^{k-1} \longrightarrow SO_p$, $i = 1, \cdots, s$ in the following way. Consider the subset $S^{k-1} \times D^{p-1} \times D^1 = S^{k-1} \times D^p \subset b(D^k \times D^p) = S^{k+j-1}$. Now, by varying the last co-ordinate t in (x,y,t) contained $S^{k-1} \times D^{p-1} \times D^1$, one can find s disjoint copies $S_i^{k-1} \times D_i^p$, $i = 1, \cdots, s$ of $S^{k-1} \times D^p$ in S^{k+p-1}. Take s copies of the $k+p$

dimensional disk, denoted by $\overline{D}_i^k \times \overline{D}_i^p$, and define maps

$\overline{s}_i : \overline{S}_i^{k-1} \times \overline{D}_i^p \longrightarrow S_i^{k-1} \times D_i^p \subset bD^{k+p}$ by $\overline{s}_i(x,y) = (x, s_i^{-1}(x)y)$.

This space is $C(k,p)$. Under the suspension of SO_p into SO_{p+1}, one obtains maps, $Ss_i : S^{k-1} \longrightarrow SO_{p+1}$. Let $C(k,p+1)$ be the manifold obtained by attaching handles using the maps Ss_i. An obvious homeomorphism can be constructed between $C(k,p) \times D^1$ and $C(k,p+1)$. Take $C(k,p) \times D^1$ and smash its boundary to a point to get a space $C(k,p) \times D^1/b(C(k,p) \times D^1)$ which is homeomorphic to $C(k,p+1)/bC(k,p+1)$. Now the boundry of $C(k,p) \times D^1$ is given as $b(C(k,p) \times D^1) = (bC(k,p) \times D^1) \cup (C(k,p) \times bD^1)$. If one studies the reduced suspension $S(C(k,p)/bC(k,p))$ of the space $C(k,p)/bC(k,p)$, it is easy to see using the previous statement, $S(C(k,p)/bC(k,p))$ is homeomorphic to $C(k,p) \times D^1/(b(C(k,p) \times D^1))$. It follows $S(C(k,p)/bC(k,p))$ is homeomorphic to $C(k,p+1)/bC(p,k+1)$.

If $m_i : S^{k-1} \longrightarrow SO_{k+j}$ are the maps used to build $C^k = C(k,k+j)$, then select maps $\underline{m}_i : S^{k-1} \longrightarrow SO_p$ such that $S^{k+j-p}(\underline{m}_i) = m_i$. Then by the above $S^{k+j-p}(C(k,p)/bC(k,p))$ is homeomorphic to

$C(k,k+j)/bC(k,k+j)$. Taking N^{k+p} equal to $C(k,p)$ completes the proof.

If $\bigvee_s S_i^{k+j}$ is the wedge of s copies of the sphere S^{k+j}, then the group $[S^{2k+j-1}, \bigvee_s S_i^{k+j}]$ is isomorphic to $\bigoplus_s \pi_{k-1}^S$, the direct sum of s copies of π_{k-1}^S. Hence, with the sequence $(a(m_1),\cdots,a(m_s))$, one can associate a homotopy class (a) belonging to $[S^{2k+j-1}, \bigvee_s S_1^{k+j}]$, where a is a representative map in this class.

Proposition 1.6. _If_ X _is equal to the space_ $D^{2k+j} \cup_a (\bigvee_s S_i^{k+j})$, _the mapping cone of the map_ a, _then_ X _is homeomorphic to_ M^{2k+j}/C_{k+j-1}.

Proof. A description of the map a along with the $a(m_i)$ will be given in the sense of Kervaire [9]. Take s copies of the 2k+j dimensional disk $\bar{D}_i^{2k+j}$ and form the connected sum of the disks along the north pole of $\bar{D}_1^{2k+j}$ with the south pole of $\bar{D}_2^{2k+j}$, then the south pole of $\bar{D}_3^{2k+j}$ with the north pole of $\bar{D}_2^{2k+j}$, etc.. The resulting manifold is a disk D^{2k+j}, the boundary being the connected sum of the spheres $\bar{S}_i^{2k+j-1}$. Each disk $\bar{D}_i^{2k+j}$ will be considered as a disk in D^{2k+j}. Let m_i represent a map $m_i:S^{k-1} \longrightarrow SO_{k+j}$ as well as its homotopy class. Define a map $m:S^{2k+j-1} \longrightarrow \bigvee_s S_i^{k+j}$ in the following way. Identify $\bigvee_s S_i^{k+j}$ with the set $\bigvee_s D_i^{k+j}/S_i^{k+j}$. Write $\bar{S}_i^{2k+k-1} = b\bar{D}_i^{2k+j}$ as $(\bar{S}_i^{k-1} \times \bar{D}_i^{k+j}) \cup (\bar{D}_i^{k} \times \bar{S}_i^{k+j-1})$. Each m_i gives rise to a map $\bar{m}_i:\bar{S}_i^{k-1} \times \bar{D}_i^{k+j} \longrightarrow S_i^{k+j}$ by $\bar{m}(x,y) = p_i(m_i(x)y)$ where $p_i:D_i^{k+j} \longrightarrow D_i^{k+k}/S_i^{k+j-1}$ denotes the projection. These maps extend to maps $m_i':\bar{S}_i^{2k+j-1} \longrightarrow D_i^{k+j}/S_i^{k+j-1}$ by m_i' restricted to $\bar{S}_i^{k-1} \times \bar{D}_i^{k+j}$ is the map $\bar{m}_i$ and sends $D_i^k \times S_i^{k+j-1}$ to the point $[S^{k+j-1}] \in D_i^{k+j}/S_i^{k+j-1}$. Up to sign, the homotopy class of m_i' equals $J_{k-1}(m_i) = a(m_i)$. By considering $\bar{D}_i^{2k+j}$ as being in D^{2k+j}, all the mpas m_i' combine to give a map $m:S^{2k+k-1} \longrightarrow \bigvee_s D_i^{k+j}/S_i^{k+j-1}$, since they agree at the points of common intersection. The homotopy class of m is the same as a up to sign.

Consider $(D_i^k \times D_i^{k+j})/(D_i^k \times S_i^{k+j-1})$ and the subset

$(0 \times D_i^{k+j})/(0 \times S_i^{k+j-1})$, which is a strong deformation retract of $(D_i^k \times D_i^{k+j})/(D_i^k \times S_i^{k+j-1})$. Let $m_i : S^{k-1} \longrightarrow SO_{k+j}$ be as before and define a map $\bar{n}_i : \bar{S}_i^{k-1} \times \bar{D}_i^{k+j} \longrightarrow (D_i^k \times D_i^{k+j})/(D_i^k \times S_i^{k+j-1})$ by $\bar{n}_i(x,y) = q_i(x, m_i(x)y)$ where q_i is the projection of $D_i^k \times D_i^{k+j}$ into $(D_i^k \times D_i^{k+j})/(D_i^k \times S_i^{k+j-1})$. Then $\bar{n}_i$ extends to a map $n_i' : \bar{S}_i^{2+j-1} \longrightarrow (D_i^k \times D_i^{k+j})/(D_i^k \times S_i^{k+j-1})$ by n_i' restricted to $S_i^{k-1} \times D_i^{k+j}$ is the map $\bar{n}_i$ and n_i' restricted to $D_i^k \times S_i^{k+j-1}$ sends this subset to the point $[D_i^k \times S_i^{k+j-1}] \in (D_i^k \times D_i^{k+j})/(D_i^k \times S_i^{k+j-1})$.

Construct a space $\bar{X} = (\bigvee_s (D_i^k \times D_i^{k+j})/(D_i^k \times S_i^{k+j-1})) \cup_n D^{2k+j}$ by identifying each point (x,y) belonging to $\bar{S}_i^{k-1} \times \bar{D}_i^{k+j}$, with $(x, m_i(x)y)$ in $(D_i^k \times D_i^{k+j})/(D_i^k \times S_i^{k+j-1})$ and sending all other points of $\bar{S}^{2k+j-1}$ into the base point of $\bigvee_s (D_i^k \times D_i^{k+j})/(D_i^k \times S_i^{k+j-1})$. Or in other words, one takes s copies of $D_i^k \times D_i^{k+j}$ and identifies each point $(x,y) \in \bar{S}_i^{k-1} \times \bar{D}_i^{k+j} \subset \bar{D}_i^{2k+j} \subset D^{2k+j}$ with its image $(x, m_i(x)y)$ in $S_i^{k-1} \times D_i^{k+j} \subset D_i^k \times D_i^{k+j}$ and the subset

$((\bigvee_s \overline{D}_i^k \times \overline{S}_i^{k+j-1})) \cup (\bigvee_s D_i^k \times S_i^{k+j-1}))$ is collapsed to a point. From the above statements, $\overline{X}$ is homeomorphic to $\bigvee_s S_i^{k+j} \cup_a D^{2k+j}$. But, $\overline{X}$ is the space obtained by attaching s handles of index k to D^{2k+j} and then identifying the boundary to a point which is precisely M^{2k+j}/C_{k+j-1}.

__Theorem 1.7.__ __For the manifold__ M^{2k+j}, __there exists the following exact sequence valid for__ $p > k+1$.

$$[\bigvee_s S_i^{k+j+1}, S^p] \xrightarrow{Sa^*} [S^{2k+j}, S^p] \xrightarrow{q^*} [M^{2k+j}, S^p] \xrightarrow{i^*}$$

$$[\bigvee_s S_i^{k+j}, S^p] \xrightarrow{a^*} [S^{2k+j-1}, S^p]$$

__where__ a^* __and__ Sa^* __are the homomorphisms given by composition with the maps__ a __and__ Sa.

__Proof.__ In C^k there are the belt disks D_i^{k+j}, $i = 1,\cdots,s$, which in C^k/bC^k become spheres S_i^{k+j}. There exists the following cofibration.

$$\bigvee_s S_i^{k+j} \xrightarrow{i} C^k/bC^k \xrightarrow{q} S^{2k+j}.$$

For X equal to $(\bigvee_s S_i^{k+j}) \cup_a D^{2k+j}$, one has the mapping cone sequence:

$$S^{2k+j-1} \xrightarrow{a} \bigvee_s S_i^{k+j} \xrightarrow{i'} X \xrightarrow{q'} S^{2k+j} .$$

By Proposition 1.6, there is a homeomorphism h from X to C^k/bC^k. h gives rise to maps such that the following diagram is commutative.

$$\begin{array}{ccccccc} & & \bigvee_s S_i^{k+j} & \xrightarrow{i} & C^k/bC^k & \xrightarrow{q} & S^{2k+j} \\ & & \downarrow & & \downarrow & & \downarrow \\ S^{2k+j-1} & \xrightarrow{a} & \bigvee_s S_i^{k+j} & \xrightarrow{i'} & X & \xrightarrow{q'} & S^{2k+j} \end{array}$$

From this it follows that the Puppe sequence,

$$\bigvee_s S_i^{k+j} \xrightarrow{i} C^k/bC^k \xrightarrow{q} S^{2k+j} \xrightarrow{t} \bigvee_s S_i^{k+j+1} \xrightarrow{Si} S(C^k/bC^k) \cdots$$

is identical from the second term on with the extended mapping cone sequence,

$$S^{2k+j-1} \xrightarrow{a} \bigvee_s S^{k+j} \xrightarrow{i'} X \xrightarrow{q'} S^{2k+j} \xrightarrow{Sa} \bigvee_s S^{k+j+1} \xrightarrow{Si'} SX \cdots$$

In particular one has that t is equal to Sa, the suspension of a, and likewise $hi = i'$, $q = q'h$. Applying the functor $[\ \ ,S^p]$, gives the exact sequence:

$$[\bigvee_s S_i^{k+j+1}, S^p] \xrightarrow{Sa^*} [S^{2k+j}, S^p] \xrightarrow{q^*} [C^k/bC^k, S^p] \xrightarrow{i^*}$$

$$\longrightarrow [\bigvee_s S_i^{k+j}, S^p] \xrightarrow{a^*} [S^{2k+j-1}, S^p]$$

Proposition 1.1 allows one to replace $[C^k/bC^k, S^p]$ in the above sequence by $[M^{2k+j}, S^p]$ provided $p > k+1$.

Proposition 1.8. *Given any subgroup* G_{k-1} *in* π_{k-1}^S, *then there exists a closed piecewise linear manifold* M^{2k+j} *such that* $[M^{2k+j}, S^{k+j+1}]$ *is isomorphic to* π_{k-1}^S/G_{k-1}, *where if* $k-1 = 2^m - 2$ *for some integer* $m > 0$, *assume* $G_{k-1} \otimes Z_2 = 0$. *If* G_{k-1} *lies in* $\operatorname{im} J_{k-1}$, *then* M^{2k+j} *may be taken to be a smooth manifold.*

Proof. For the definitions and assertions concerning PL, see the Appendix. The map:

$$J_{k-1}^{\overline{PL}}: \pi_{k-1}(\overline{PL}_{k+j}(D)) \longrightarrow \pi_{k-1}^S$$

is an onto map if $k-1$ is not of the form $2^m - 2$ for some m. If $k-1 = 2^m - 2$, then $J_{k-1}^{\overline{PL}}$ is a map which is onto the odd torsion elements of π_{k-1}^S. Since π_{k-1}^S is a finite group, there exist piecewise linear maps, $f_i : S_i^{k-1} \times D_i^{k+j} \longrightarrow S^{k-1} \times D^{k+j}$, $i = 1,2,\cdots,s$,

such that the following is true. If (f_i) is the homotopy class of

f_i in $\pi_{k-1}(\overline{PL}_{k+j}(D))$ and $H_{k-1} = ((f_1),(f_2),\cdots,(f_s))$ is the sub-

group of $\pi_{k-1}(\overline{PL}_{k+j}(D))$ generated by the (f_i), then

$J_{k-1}^{\overline{PL}}(H_{k-1}) = G_{k-1}$. Construct a piecewise linear C^k, by glueing

handles, $D_i^k \times D_i^{k+j}$ to $S^{2k+k-1} \subset D^{2k+j}$, using the maps f_i. Let

M^{2k+j} be the double of C^k, which is a closed PL manifold. From

Proposition 1.1, there exists the following exact sequence:

$$\bigoplus_s Z_i \xrightarrow{t^*} [S^{2k+j}, S^{k+j+1}] \xrightarrow{q^*} [M^{2k+j}, S^{k+j+1}] \longrightarrow 0 .$$

As shown in the proof of Theorem 1.7, the map t given by 1.1

coincides with the map Sa. The claim is the image of

$t^* : \bigoplus_s Z_i \longrightarrow [S^{2k+j}m\ S^{k+j}]$ is the subgroup generated by all the

$a(m_i)$. Indeed, $t^* = Sa^*$ is given by $S^*a(b) = Sa \circ b$ and for two

maps, $f : S^{2k+j} \longrightarrow \bigvee_s S_i^{k+j+1}$, and $g : \bigvee_s S^{k+j+1} \longrightarrow S^{k+j+1}$, the homotopy class $(g \circ h)$ is equal to the homotopy class $\sum^{s} (g_i \circ f)$ where g is equal to $\bigvee_s g_i$. (A quick proof can be given by using the Thom-Pontriagin interpretation of $g \circ f$ and noting that the dimension conditions are such that none of the manifolds are linked.) Choosing the maps $(0, \cdots, i_j, \cdots, 0)$, $j = 1, \cdots, s$, in $[\bigvee_s S^{k+j+1}, S^{k+j+1}]$, where $i_j : S_j^{k+j+1} \longrightarrow S_j^{k+j+1}$ is the identity map, the assertion follows.

The second statement is established by noting that if G_{k-1} is contained in $\operatorname{im} J_{k-1}$, then one may choose the maps $f_i : S_i^{k-1} \times D_i^{k+j} \longrightarrow S^{k-1} \times D^{k+j}$ such that the f_i are differentiable maps.

<u>Remark</u>. The proposition states that if one takes $G_{k-1} = \operatorname{im} J_{k-1}$, then there is a smooth manifold such that $[M^{2k+j}, S^{k+j+1}]$ is isomorphic to coker J_{k-1}. If one takes $G_{k-1} = \operatorname{coker} J_{k-1}$, then one can only assert there is a piecewise linear M^{2k+j} such that

$[M^{2k+j}, S^{k+j+1}]$ is isomorphic to im J_{k-1}.

If M^{2n} is an $n-1$ connected smooth manifold with $n \geq 3$, then one can derive elements $n_i \in \pi_{n-1}(SO_n)$ similar to the way the m_i were previously derived. By the suspension S of $SO_n \longrightarrow SO_{n+1}$, one obtains elements $Sn_i \in \pi_{n-1}(SO)$ and applying the stable Hopf-Whitehead homomorphism J_{n-1} to the Sn_i, one gets a sequence $a(Sn_i) \in \pi^S_{n-1}$, $i = 1,\cdots,s$, where s is the rank of $H^n(M^{2n}, Z)$.

Corollary 1.9. *If (a) denotes the subgroup generated by* $(a(Sn_1),\cdots,a(Sn_s))$ *in* π^S_{n-1}, *then* $[M^{2n}, S^{n+1}]$ *is isomorphic to* $\pi^S_{n-1}/(a)$.

Proof. Because C^n/bC^n is homeomorphic to M^{2n}, $[C^n/bC^n, S^{n+1}]$ is isomorphic to $[M^{2n}, S^{n+1}]$. For the moment, let the notation of the technical Lemma 1.4 be used, i.e., $C^n = C(n,n)$. Since one is dealing with homotopy classes in the stable range, the suspension gives an isomorphism of $[C(n,n)/bC(n,n), S^{n+1}]$ with the group $[S(C(n,n)/bC(n,n)), S^{n+2}]$. However, the proof of the technical lemma gives that $S(C(n,n)/bC(n,n))$ is homeomorphic to $C(n,n+1)/bC(n,n+1)$. But by 1.8, $[C(n,n+1)/bC(n,n+1), S^{n+2}]$ is isomorphic to $\pi_{n-1}^S/(a)$.

The next proposition gives a complete analysis of the case when $p = k+j$. Assume $k > 2$, $j \geq 2$.

Proposition 1.10. For the manifold M^{2k+j} with the associated sequence $(m_1, \cdots, m_s)$, one has the following split exact sequences:

(i) If when the m_i's are reduced modulo 2, some m_i stays non-zero, then one has the sequence,

$$0 \longrightarrow \operatorname{coker} J_k \longrightarrow [M^{2k+j}, S^{k+j}] \longrightarrow H^k(M^{2k+j}, Z) \longrightarrow 0$$

(ii) In all other cases,

$$0 \longrightarrow \pi_k^S \longrightarrow [M^{2k+j}, S^{k+j}] \longrightarrow H^k(M^{2k+j}, Z) \longrightarrow 0$$

Proof. By 1.7, one has the exact sequence

$$[\bigvee_s S_i^{k+j+1}, S^{k+j}] \xrightarrow{Sa*} [S^{2k+j}, S^{k+j}] \xrightarrow{q*} [M^{2k+j}, S^{k+j}] \xrightarrow{i*}$$

$$\longrightarrow [\bigvee_s S_i^{k+j}, S^{k+j}] \xrightarrow{a*} [S^{2k+j-1}, S^{k+j}].$$

Note $i*$ is not the zero map, for $[S^{2k+j-1}, S^{k+j}]$ is a stable group. More precisely, $i*$ is a C isomorphism in the sense of Serre, where C is the class of finite abelian groups. Therefore, the rank of the image of i^* is equal to s. Hence, there is an isomorphism (non-canonical) of the image of i^* with $H^k(M^{2k+j}, Z)$. Consequently, the sequence becomes

$$\bigoplus_s (Z_2)_i \xrightarrow{Sa^*} [S^{2k+j}, S^{k+j}] \xrightarrow{q^*} [M^{2k+j}, S^{k+j}] \longrightarrow H^k(M^{2k+j}, Z) \longrightarrow 0$$

<u>Case 1</u>. $k \equiv 0 \mod (8)$. In this case $\operatorname{im} J_k$ is equal to Z_2 and Sa^* is mapped into $\operatorname{im} J_k$. The latter part of this last statement is due to Novikov (see 1.11). If some m_i is odd, it follows from statement 1.11, Sa^* is onto $\operatorname{im} J_k$, while if all m_i are even, Sa^* is the zero map.

<u>Case 2</u>. $k \equiv 1 \mod (8)$. In this case again $\operatorname{im} J_k$ is Z_2 and Sa^* is mapped into $\operatorname{im} J_k$. If some m_i in $(m_1, \cdots, m_s)$ is non-zero, by statement 1.11, Sa^* is an onto map and if all m_i are zero, Sa^* is the zero map.

<u>Case 3</u>. If $k \equiv 2, 3, 4, 5, 6, 7 \mod (8)$, then Sa^* is the zero map. For $k \equiv 3, 5, 6, 7, \mod (8)$, all m_i are zero and therefore, Sa^* is trivially the zero map. If $k \equiv 2, 4 \mod (8)$, then Sa^* has its image contained in $\operatorname{im} J_k$. This last group though, is zero for $k > 2$.

This concludes the proof of Proposition 1.10 .

The next assertion takes care of the case when $k = 2$, since a certain pecularity arises.

Addendum 1.10. *If* $k = 2$, $j \geq 2$, *and some* m_i *in the sequence* $(m_1, m_2, \cdots, m_s)$ *associated with* M^{2k+j} *is odd, then* $[M^{2k+j}, S^{k+j}]$ *is isomorphic to* $H^k(M^{2k+j}, Z)$ *otherwise,* $[M^{2k+j}, S^{k+j}]$ *is isomorphic to* $Z_2 \oplus H^k(M^{2k+j}, Z)$.

Proof. From Proposition 1.7, one derives the following exact sequence.

$$\bigoplus_s Z_2 \xrightarrow{Sa*} \pi_k^S \longrightarrow [M^{2k+j}, S^{k+j}] \longrightarrow H^k(M^{2k+j}, Z) \longrightarrow 0$$

For $k = 2$, $\pi_k^S = Z_2$, and by statement 1.11, if some m_i is odd, $Sa*$ is onto π_k^S, (contrary to the case when $k > 2$), while if all the m_i are even, $Sa*$ is the zero map.

The following statement can be found in [11].

Statement 1.11. If $a \in \text{im } J_k$ and $b \in \pi_n^S$, $k \geq n$, then the composition $a \circ b \in \text{im } J_{k+n}$, except for $k = n = 1, 3, 7$. For the non-zero element $\alpha = J_1(1)$ in π_1^S, then α^2 is a generator of π_2^S. For $s \geq 1$, $\alpha \circ J_{8s-1}(1) = J_{8s}(1)$, and $\alpha^2 \circ J_{8s-1}(1) = J_{8s+1}(1)$.

Section 2. Determination of the Obstruction. In this section, M^{2k+j} means a smooth manifold of the type described after Corollary 1.2 with $j, k \geq 2$. For the group $[M^{2k+j}, S^{k+j}]$, a framed manifold (N^k, F^{k+j}) in M^{2k+j} representing a cohomotopy class can have a non-trivial intersection with a belt disk of M^{2k+j}. One purpose of this section is to give a description of the possible values this intersection number can take. Indeed, it will be shown that the intersection number cannot take arbitrary values. Another purpose of this section is to find if a given cohomotopy class has a representative which has a particular nice form such as a homotopy sphere, and if it cannot be taken as a homotopy sphere, what is the obstruction, calculation of the obstruction and proving this obstruction is realizable.

Surgery Lemma 2.1. Let M^{p+k} be a $k-1$ connected smooth manifold and (N^t, F^{p+k-t}) a framed submanifold of M^{p+k}. Assume the following conditions are satisfied:

(i) $[t/2] < k$,

(ii) $[t/2] + 1 < p+k-t$, where $[t/2]$ is as before.

Then one can frame cobord (N^t, F^{p+k-t}) in M^{p+k} such that the new framed manifold is $[t/2]$ connected if t is odd and $[t/] - 1$

connected if t *is even. In case* t *is twice an odd number which is not of the form* $2^m - 2$, *the new framed manifold may be taken to be a homotopy sphere. For other even values of* t, *a surgery obstruction arises to obtaining a manifold which is* $[t/2]$ *connected. The obstruction is given by the index if* t *is divisible by* 4 *and the Kervaire invariant if* t *is of the form* $2^m - 2$ *for some* $m > 0$. *If the obstruction is zero, then the new framed manifold may be taken to be a homotopy sphere provided* $t \geq 5$.

Proof. The procedures to be used are those given in [5].

Let $f:S^r \longrightarrow N^t$ be a map with $r \leq t/2$. (If $r = 2$, assume $t \geq 5$.) Then by [4], f is homotopic to a map $e:S^r \longrightarrow N^t$ which is an embedding. Since $r \leq [t/2]$, condition (i) and the connectivity of M^{p+t} together imply this map may be extended to an embedding $\bar{e}:D^{r+1} \longrightarrow M^{p+t}$. Condition (ii) permits one to assume that the intersection of N^t with the image of D^{r+1} by the map $\bar{e}$, consists only of $\bar{e}(S^r)$.

The above shows that the methods of [10], [12] may be employed, whence the conclusion of the lemma using also the improvement given in

[3].

A modified topological or piecewise linear version of the above lemma is also true.

By putting Proposition 1.8 and Lemma 2.1 together, one may state the following corollary.

Corollary 2.2. *Each homotopy class* α *in* $[M^{2k+j}, S^{k+j+1}]$ *has a representative which can be taken to be a framed homotopy sphere provided* $k \geq 5$ *is not of the form* $2^m - 2$ *for some positive* m.

Interpreting a cohomotopy class as a framed cobordism class, consider the obstruction which arises when the framed manifold has intersections with the belt disks which consists of a certain number of points and for each belt disk the algebraic sum is non-zero. Now, for each framed manifold by Lemma 2.1, there is also a surgery obstruction to finding a representative in a framed cobordism class which is a homotopy sphere. For the case when k is congruent to 0 modulo 4, there is an intimate connection between these two obstructions as the next result shows.

The following notation will be needed. Suppose x^k and y^{n-k} are two homology classes of complementary dimensions in a manifold, then

$KI(x^k, y^{n-k})$ will denote the Kronecker intersection of these two classes. Consider the power series (see [6]) $\sqrt{z}\coth\sqrt{z}$ and the associated multiplicative sequence $L_r(x_1,x_2,\cdots,x_r)$. Set s_r equal to the coefficient of x_r in $L_r(x_1,x_2,\cdots,x_r)$ which is equal to $2^{2r}(2^{2r-1}-1)B_r/(2r!)$ where B_r is the r-th Bernoulli number. a_r is equal to 1 or 2 depending whether r is even or odd.

Theorem 2.3. *Let M^{2k+j} be the manifold as before with associated sequence $(m_1,\cdots,m_s)$ and $k = 4r > 0$. If (N^k, F^{k+j}) is a framed manifold corresponding to a map $f:M^{2k+j} \longrightarrow S^{k+j}$, then the index $I(N^k)$ of N^k is equal to*

$$I(N^k) = s_r a_r (2r-1)!\left(\sum^s m_i KI(N^k, z_i^{k+j})\right)$$

where z_i^{k+j}, $i = 1,2,\cdots,s$ is the homology class in $H_{k+j}(M^{2k+j}, Z)$ obtained by extending the belt disk to a closed cycle in M^{2k+j}.

Corollary 2.4. *(N^k, F^{k+j}) is framed cobordant in M^{2k+j} to a framed homotopy sphere if and only if*

$$\sum^s m_i KI(N^k, z_i^{k+j}) = 0$$

and $k = 4r > 4$.

The corollary follow immediately from the Surgery Lemma 2.1.

Proof. Set g equal to the integral part of $j/4$. Denote by $P(T(N^k)) = \sum^r q_i$, $q_i \in H^{4i}(N^k, Z)$ and $P(T(M^{2k+j})) = \sum^{r+g} p_i$, $p_i \in H^{4i}(M^{2k+j}, Z)$, the total Pontriagin class of the tangent bundles of N^k and M^{2k+j} respectively. The the index of N^k is given by

$$I(N^k) = (L_r(q_1,q_2,\cdots,q_r), N^k)$$

where N^k also represents the fundamental class in $H_k(N^k, Z)$. It is well known that the Pontriagin class of a Whitney sum of two vector bundles E_1, E_2 is equal to the product of the individual Pontriagin classes of the vector bundles modulo 2 torsion. From this and the naturality of the Pontriagin classes, $q_i = i^*p_i$ where i^* is the induced map of cohomology given by the inclusion $i:N^k \longrightarrow M^{2k+j}$.

Hence,

$$I(N^k) = (L_r(p_1,p_2,\cdots,p_r), i_*(N^k)).$$

Since $H^i(M^{2k+j}, Z) = 0$ for $i < k$, it follows $q_i = 0$ for $i < k$, and therefore,

$$I(N^k) = (s_r p_r, i_*(N^k)).$$

From the structure of M^{2k+j}, $H^k(M^{2k+j}, Z) = \bigoplus_s Z_i$ with a preferred set of generators $z_1^*, z_2^*, \cdots, z_s^*$ which are the HOM duals of homology classes $z_1, z_2, \cdots, z_s$ given by extending the core disks to embedded spheres in M^{2k+j}. With this, one may write the r-th Pontriagin class as $\sum^s \bar{p}_i z_i^*$ where the $\bar{p}_i$ are integers. Denote by z_i^{k+j} in $H_{k+j}(M^{2k+j}, Z)$, the Poincare dual of z_i^*. Then $KI(N^k, z_i^{k+j})$ represents the intersection of N^k with the belt disk D_i^{k+j} of the i-th handle. Hence, $i_*(N^k) = \sum^s KI(N^k, z_i^{k+j}) z_i$ and therefore, $I(N^k) = \sum^s s_r(\bar{p}_i KI(N^k, z_i^{k+j}))$.

<u>Sub-lemma 2.5</u>. <u>The integer</u> $\bar{p}_i$, <u>is equal to</u> $(2r-1)! a_r m_i$.

<u>Proof</u>. Let C_i^k be the submanifold of M^{2k+j} obtained by neglecting all handles of index k in C^k except for the i-th handle. Then C_i^k is a smooth manifold with boundary and $H^k(C_i^k, Z)$ is isomorphic to Z such that under the inclusion of C_i^k into M^{2k+j}, the cohomology group $H^k(M^{2k+j}, Z) = \bigoplus_s Z_i$ projects onto the i-th factor $Z_i = H^k(C_i^k, Z)$ and carries the generator z_i^* onto a generator $\bar{z}_i^*$ of $H^k(C_i^k, Z)$. If $T(C_i^k)$ denotes the tangent bundle of C_i^k, and

$i_i : C_i^k \longrightarrow M^{2k+j}$ the inclusion, then

$$P(T(C_i^k)) = P\left(i_i^! T\left(M^{2k+j}\right)\right) = 1 + \bar{p}_i \bar{z}_i^* .$$

Now C_i^k can be interpreted as the total space of a $k+j$ dimensional disk bundle E_i^{k+j} over S^k. Denote by b_i the projection map. Since the total Pontriagin class of S^k is equal to 1, it follows,

$$P(T(C_i^k)) = b_i^*(P(E_i^{k+j})).$$

The following statement can be found in [8].

If E^p is a disk bundle over S^k, $k = 4r$ $r > 0$, then the r-th Pontriagin class $p_r(E^p)$ is related to its characteristic class $x(E^p) \in \pi_{k-1}(SO_p)$ by

$$p_r(E^p) = (2r-1)! a_r m z^k$$

where (1) z^k is a generator of $H^k(S^k, Z)$, (2) m is the integer in $\pi_{k-1}(SO)$ to which $x(E^p)$, through interated suspension maps to, and the suspension is $SO_p \longrightarrow SO$, (3) a_r is 1 or 2 depending upon whether r is even or odd.

Since $j > 1$, E_i^{k+j} is stable, and therefore, the integer $\bar{p}_i$ is equal to $(2r-1)! a_r m_i$. This completes the proof of the sub-lemma.

One can rewrite the equation for the index using the above lemma as

$$I(N^k) = s_r a_r (2r-1)!\left(\sum^{s} m_i KI\left(N^k, z_i^{k+j}\right)\right).$$

Unfortunately, the proposition is true only in the smooth case. However, in the topological case, one can use it to motivate certain definitions. (see Appendix).

The next proposition gives some insight into the tangental structure of the manifold $\left(N^k, F^{k+j}\right)$.

Proposition 2.6. *If $\left(N^k, F^{k+j}\right)$ is a framed manifold in M^{2k+j} corresponding to a map $f: M^{2k+j} \longrightarrow S^{k+j}$, then N^k is an almost parallelizable manifold for arbitrary $k \geq 2$.*

By definition N^k is almost parallelizable if after the removal of the interior of a finite number of disks $\mathring{D}_1^k, \cdots, \mathring{D}_q^k$, the resulting manifold is parallelizable. In [10], it is proved that a manifold with non-empty boundary is parallelizable if and only if the manifold has a trivial stable normal bundle. One may assume $\left(N^k, F^{k+j}\right)$ intersects the belt disks D_i^{k+j}, $i = 1,2,\cdots,s$, transversally. Take a small tubular neighborhood of the belt disk D_i^{k+j} in C^k and denote it by

$\mathring{D}_i^k \times D_i^{k+j}$. Then $C^k - (\bigcup_s (\mathring{D}_i^k \times D_i^{k+j}))$ is diffeomorphic to the disk D^{2k+j}, and $N^k - (\bigcup_s (\mathring{D}_i^k \times D_i^{k+j}))$ is a manifold with a finite number of open disks removed. Since $N^k - (\bigcup_s (\mathring{D}_i^k \times D_i^{k+j}))$ has the frame $\bar{F}^{k+j}$ given by restricting the frame F^{k+j} to $N^k - (\bigcup_s (\mathring{D}_i^k \times D_i^{k+j}))$, one has $N^k - (\bigcup_s (\mathring{D}_i^k \times D_i^{k+j}))$ imbedded in D^{2k+j} with trivial normal bundle. Consequently, N^k is almost parallelizable.

Corollary 2.7. *If* $k = 4r > 0$, *then every cohomotopy class in* $[M^{2k+j}, S^{k+j}]$ *contains as a representative, a* $2r-1$ *connected almost parallelizable manifold whose associated quadratic form is of type* II.

Proof. By the Surgery Lemma 2.1 and Proposition 2.6 above, one may always find a $2r-1$ connected almost parallelizable N^k. Recall, that for such N^k the quadratic form given by the intersection pairing $(\ ,\ )$ of $H_{2r}(N^k, Z)$ is of type II if for each x^{2r} belonging to $H^{2r}(N^k, Z_2)$, $Sq^{2r}(x^{2r})$ is zero where Sq^{2r} is the Steenrod opertion. However, if N^k is a $2r-1$ connected manifold, then $(\ ,\ )$ is of type II if and only if the Wu class U_{2r} is zero. Since N^k is almost parallelizable, the corollary follows.

<u>Remark</u>. Milnor and Kervaire [10] analyze for almost parallelizable closed manifolds N^{4r} a number I_r which is the "smallest" positive index occurring among such manifolds. Moreover, the index of any closed almost parallelizable manifold is some multiple of it. It is given by, $I_r = s_r a_r (2r-1)! j_{k-1}$ where j_{k-1} denotes the order of im J_{k-1} in π^S_{k-1} and the other notation is as before. Now, if (N^k, F^{k+j}), $k = 4r$, is a representative of a homotopy class in $[M^{2k+j}, S^{k+j}]$, by 2.3, 2.6, and the above, there is an integer $p(N^k)$ such that

$$p(N^k) j_{k-1} = \sum^s m_i KI(N^k, z_i^{k+j}).$$

Hence, for a fixed set $(m_1, \cdots, m_s)$ the intersections $KI(N^k, z_i^{k+j})$ of N^k with the belt disks in M^{2k+j} cannot be arbitrary. In Section 3, $p(N^k)$ will be calculated for some special manifolds.

2.3 and 2.4 examine the cases when $k \equiv 0 \mod (4)$. The only other time when the sequence $(m_1, \cdots, m_s)$ associated with M^{2k+j} will have a non-zero element is if $k \equiv 1,2 \mod (8)$. In these cases a much sharper result than 2.4, can be stated.

<u>Corollary 2.8</u>. <u>If</u> $k \equiv 1,2$ <u>mod</u> (8), $k \neq 2$, <u>then every homotopy class in</u> $[M^{2k+j}, S^{k+j}]$ <u>has a representative which can be taken to be a framed homotopy sphere</u>. <u>If</u> $k = 2$ <u>and some</u> m_i <u>is not zero, the conclusion is still true</u>.

<u>Proof</u>. $k \geq 3$, the Surgery Lemma applies.

If $k = 2$, then the obstruction is the Kervaire invariant. It will be shown that each framed cobordism class has a representative manifold for which the Kervaire invariant is zero. Consider the following part of the exact sequence:

$$[\bigvee_s S_i^{3+j}, S^{2+j}] \xrightarrow{Sa^*} [S^{4+j}, S^{2+j}] \xrightarrow{q^*} [M^{4+j}, S^{2+j}]$$

Sa equals $\bigvee_s Sa(m_i)$, and the map Sa^* is given by composition of Sa with elements of $\left(\bigvee_s S_i^{3+j}, S^{2+j}\right) = \bigoplus_s \{\pi_1^S\}_i$. Suppose some m_i is not zero, say m_1. Then $a(m_1) = J_1(1)$, which is a generator of π_1^S. Denote this generator by α. The element $(\alpha,0,\cdots,0)$ in $\bigoplus_s \{\pi_1^S\}_i$, gives by the above, $Sa^*(\alpha,0,\cdots,0) = \alpha^2$ which is a non-zero generator of $\left(S^{4+j}, S^{2+j}\right) = \pi_2^S$. This element α^2 has non-zero Kervaire invariant in π_2^S i.e., there is a framed manifold with non-zero Kervaire invariant representing α^2. However, the sequence is exact, which implies that α^2 thought as an element in $[M^{4+j}, S^{2+j}]$, via the map q, is zero or the manifold with non-zero Kervaire invariant in S^{4+j} bounds when considered as an element in M^{4+j}, The Kervaire invariant is of order two. If (N^2, F^{2+j}) is a framed manifold in M^{4+j} with non-zero Kervaire invariant, then adding the bounding framed manifold given above to (N^2, F^{2+j}), a new framed manifold with zero Kervaire invariant is obtained which is framed cobordant to (N^2, F^{2+j}) still.

Section 3. Sphere Bundles Over Spheres. Here the results of the previous sections are applied to stable sphere bundles over spheres. G^n will denote the total space of an n-dimensional sphere bundle over the sphere S^k with $n > k \geq 2$. If E^{n+1} is the $n+1$ disk bundle associated with G^n, then G^n may be realized as the boundary of E^{n+1} i.e., $G^n = bE^{n+1}$. Since the situation is stable, one may write $E^{n+1} = E^n \oplus \bar{D}^1$, the Whitney sum of an n-dimensional disk bundle E^n over S^k and the trivial one dimensional disk bundle $\bar{D}^1 = S^k \times D^1$. Taking two copies E_1^n and E_2^n of E^n and identifying them along their boundaries by the identity map gives the sphere bundle G^n.

Let $x(E^n) \in \pi_{k-1}(SO_n)$ be the characteristic element of the bundle of E^n. Then E^n is constructed from $x(E^n)$ as follows. Take two copies D_1^{k+n}, D_2^{k+n} of the $k+n$-dimensional disk and look at their boundaries:

$$bD_i^{k+n} = b\left(D_i^k \times D_i^n\right) = \left(S_i^{k-1} \times D_i^n\right) \cup \left(D_i^k \times S_i^{n-1}\right)$$

$i = 1, 2$. Identifying the subset $S_2^{k-1} \times D_2^n$ with $S_1^{k-1} \times D_1^n$ by the map $F(z,y) = (z, x(E^n)(z)y)$ gives the bundle E^n over S^k. But, this is the same as attaching a handle of index k to the $k+n$ dimensional disk.

Using handle-body theory of the previous sections, G^n is a smooth manifold which admits a Smale-Wallace function with the following data: the only critical values are of index $0, k, n-k, n$ and for each critical value there is only one critical point in G^n corresponding to this value.

The disk bundle E^n is equal to C^k in the earlier notation, and $x(E^n)$ is identical with the number m_1. Hence, the sequence associated with the manifold G^n is given by $(x(E^n))$.

By Proposition 1.1, the homotopy classes of maps $[G^n, S^p]$ is isomorphic to $[E^n/bE^n, S^p]$ for $p > k+1$. But, E^n/bE^n is by definition the Thom space $\Psi(E^n)$ of the bundle E^n, and $[E^n/bE^n, S^p]$ can be interpreted as the cobordism classes of framed manifolds in E^n which do not touch the boundary. Moreover, $[E^n/bE^n, S^p]$ always has a natural abelian group structure. (If $k = 2, 4, 8$, assume $n \geq k+2$.) More generally, one has the following.

__Proposition 3.1.__ *Suppose a q-dimensional disk bundle E^q over S^k is equivalent to the Whitney sum $E^{q-r} \oplus \bar{D}^r$ of an $q-r$ dimensional disk bundle E^{q-r} over S^k and a trivial r-dimensional disk bundle $\bar{D}^r$ with $r \geq 1$, q arbitrary. Then $[E^q/bE^q, S^p]$ has a natural group structure, and if $r \geq 2$, this structure is abelian.*

__Proof.__ Since $E^q = E^{q-r} \oplus \bar{D}^r$, one has for the Thom spaces $E^q/bE^q = \Psi(E^q) = \Psi(E^{q-r} \oplus D^r)$. However, $\Psi(E^{q-r} \oplus D^r)$ is homeomorphic to $S^r(\Psi(E^{q-r}))$, the r-fold suspension of $\Psi(E^{q-r})$. If $r \geq 1$, then $[S^r(\Psi(E^{q-r})), S^p]$ always has a natural group structure and if $r \geq 2$, this structure is abelian.

By application of the technical Proposition 1.4 and the above for a "metastable" bundle E^n, $[E^n/bE^n, S^p]$ has a group structure subject to the restriction of 1.4.

Unless otherwise stated, assume for the rest of the section $n \geq k+2$. Then one can write n as $n = k+j$, $j \geq 2$, and E^{k+j} shall mean the $k+j$ dimensional disk bundle which was derived from the sphere bundle G^{k+j} at the beginning of this section.

<u>Proposition 3.2.</u> <u>If</u> G^{k+j} <u>is a trivial bundle over</u> S^k, <u>then for all</u> $p > k+1$, $[G^{k+j}, S^p]$ <u>is isomorphic to</u> $[S^{2k+j}, S^p] \oplus [S^{k+j}, S^p]$.

<u>Proof.</u> From 1.7, there exists the exact sequence of abelian groups for $p > k+1$.

$$[S^{k+j+1}, S^p] \xrightarrow{Sa^*} [S^{2k+j}, S^p] \xrightarrow{q^*} [G^{k+j}, S^p] \xrightarrow{i^*}$$

$$\longrightarrow [S^{k+j}, S^p] \xrightarrow{a^*} [S^{2k+j-1}, S^p]$$

where $a = J_{k-1}(x(E^{k+j})) = 0$.

Define a homomorphism $r^*: [S^{k+j}, S^p] \longrightarrow [G^{k+j}, S^p]$ as follows. Since G^{k+j} is trivial, $E^{k+j} \subset G^{k+j}$ is equal to $S^k \times D^{k+j}$. If (N^{k+j-p}, F^p) is a framed manifold in S^{k+j}, representing a homotopy class in $[S^{k+j}, S^p]$, then (N^{k+j-p}, F^p) sits in a disk D^{k+j} in S^{k+j}. The framed manifold $(S^k \times N^{k+j-p}, F^p)$ is a framed manifold in $S^k \times D^{k+j} = E^{k+j}$. Define r^* as $r^*(N^{k+j-p}, F^p) = (S^k \times N^{k+j-p}, F^p)$. It is not difficult to check that this gives a well defined homomorphism from $[S^{k+j}, S^p]$ to $[G^{k+j}, S^p]$ and i^*r^* is the identity. Since the sequence consists of abelian groups, it splits.

Remark. The proposition does not assume $2p - 2 \geq 2k + j$.

The only time when G^{k+j} would not be a trivial stable sphere bundle is for $k \equiv 0, 1, 2, 4 \mod (8)$. These cases will now be examined. Let $(Jx(G))$ denote the cyclic subgroup of π^S_{k-1} generated by $J_{k-1}(x(E^{k+j}))$; $(x(G^{k+j}))_2$ will denote the mod 2 reduction of the number $x(E^{k+j})$.

Proposition 3.3. (*Classification Proposition for stable sphere bundles.*) *Suppose* $k \equiv 0, 1, 2, 4$, *mod* (8), *and let* G^{k+j} *be a sphere bundle over* S^k. *Then*

(1) $[G^{k+j}, S^p] = \pi^S_{2k+j-p}$ for $p \geq k+j+2$

(2) $[G^{k+j}, S^{k+j+1}] = \pi^S_{k-1}/(Jx(G))$

(3a) $[G^{k+j}, S^{k+j}] = Z \oplus \pi^S_k$ if $(x(G^{k+j}))_2 = 0$

(3b) $[G^{k+j}, S^{k+j}] = Z \oplus \text{coker } J_k$ if $(x(G^{k+j}))_2 \neq 0$, $k > 2$

(3c) $[G^{2+j}, S^{2+j}] = Z$ if $k = 2$ and $(x(G^{2+j}))_2 \neq 0$

Assume $j \geq 4$. There exist the exact sequences:

(4a) $0 \longrightarrow \pi^S_{k+1} \longrightarrow [G^{k+j}, S^{k+j-1}] \longrightarrow Z_2 \longrightarrow 0$

if $(x(G^{k+j}))_2 = 0$

(4b) $0 \longrightarrow \text{coker } J_{k+1} \longrightarrow [G^{k+j}, S^{k+j-1}] \longrightarrow Z_2 \longrightarrow 0$

if $(x(G^{k+j}))_2 \neq 0$ $k > 2$

(4c) $0 \longrightarrow Z_{12} \longrightarrow [G^{2+j}, S^{j+1}] \longrightarrow Z_2 \longrightarrow 0$

if $(x(G^{2+j}))_2 \neq 0$, $k = 2$.

The proposition follows from earlier statements.

Remark. If $j = 4$, then the above gives a determination of the cohomotopy groups of a sphere bundle G^{k+4} over S^k modulo the cohomotopy groups of spheres.

For the homology group, $H_k(G^{k+j}, Z) = Z$, Serre gave an upper bound for the "best" homology class which can be realized by a framed manifold [15]. By "best" homology class realizable by a framed manifold, one means a homology class dx, x a generator of $H_k(G^{k+j}, Z)$, and d is the smallest non-zero positive integer such that dx is realizable by a framed manifold. The next few results completely determine this number d which improves Serre's result with respect to sphere bundles. The statements also give information on the form of the manifold realizing this class.

The characteristic class $x(E^{k+j})$ will be written simply as m, where m is an integer or belongs to Z_2 depending upon whether $\pi_{k-1}(SO)$ is Z or Z_2, and $\bar{m}$ shall mean the integer representing the residue class of m modulo the denominator of $B_r/4r$ with $0 \leq \bar{m} <$ (denominator $B_r/4r$). Denote by $(\bar{m}, B_r)$, the denominator of $B_r/4r$ divided by the greatest common divisor of $\bar{m}$ and the denominator of $B_r/4r$.

<u>Theorem 3.5.</u> <u>Let</u> G^{k+j} <u>be a stable</u> $k+j$ <u>sphere bundle over</u> S^k, $k = 4r > 0$, <u>classified by the integer</u> $m \neq 0$. <u>Then the best homology class in</u> $H_k(G^{k+j}, Z)$ <u>realizable by a framed submanifold is equal to</u>

$(\overline{m}, B_r)x$ *where* x *is a generator of* $H_k(G^{k+j}, Z)$.

Corollary 3.6. The index of a framed manifold realizing the homology class $(\overline{m}, B_r)x$ *is equal to the number*

$$a_r m(\overline{m}, B_r)2^{2r-1}(2^{2r-1} - 1)B_r/r.$$

Consequently, for $m \neq 0$, *the framed manifold cannot be taken to be a sphere. However, a framed manifold can be found which is* $2r-1$ *connected, almost parallelizable and whose quadratic form is of type II.*

Proof of 3.5. Let j_{k-1} denote the order of the image of J_{k-1}. Then Adams showed for $k \equiv 1, 2, 4$, im J_{k-1} is a direct summand of π^S_{k-1} and for $k \equiv 4 \mod (8)$, j_{k-1} = denominator $B_r/4r$ [1]. It follows from Adams' paper, and the recent proof of the Adams' conjecture by Quillen [13], that for $k \equiv 0 \mod (8)$ im J_{k-1} is a direct summand also and j_{k-1} = denominator $B_r/4r$.

Write $m = nj_{k-1} + \overline{m}$, then $J_{k-1}(m) = J_{k-1}(\overline{m})$. In the group im J_{k-1}, the order of the sub-group generated by $J_{k-1}(m) = J_{k-1}(\overline{m})$ is equal to denominator $B_r/4r$ divided by the greatest common divisor of $\overline{M}$ and denominator $B_r/4r$, i.e., $(\overline{m}, B_r)$.

As before, one has the exact sequence:

$$[G^{k+j}, S^{k+j}] \xrightarrow{i^*} [S^{k+j}, S^{k+j}] \xrightarrow{a^*} [S^{2k+j-1}, S^{k+j}]$$

where $a = J_{k-1}(m)$. Since $[S^{k+j}, S^{k+j}] = Z$, image of i^* equals dZ for some positive integer d. From earlier remarks, the map i^* is given as $i^*(N^k, F^{k+j}) = KI(N^k, z^{k+j})$, where z^{k+j} is a generator of $H_{k+j}(G^{k+j}, Z)$ given by a fibre in the sphere bundle G^{k+j}, and (N^k, F^{k+j}) is a framed manifold in a framed cobordism class corresponding to some cohomotopy class. Hence, the best homology class (non-zero) in $H_k(G^{k+j}, Z)$ is dx, x a generator of $H_k(G^{k+j}, Z)$. But, by the exactness of the sequence above $\operatorname{im} i^* = \ker J_{k-1}(m)$ which by the previous statements is generated by $(\bar{m}, B_r)$.

The corollary follows from 2.3 and 2.7. For any other framed manifold (V^k, F^{k+j}) by the above, $KI(V^k, Z^{k+j})$ must be some integral multiple of $(\bar{m}, B_r)$. Therefore, for sphere bundles G^{k+j}, the number given in the remark after 2.7 has been determined.

The corollary shows for $k > 4$, $m \neq 0$, a homotopy class α in $[G^{k+j}, S^{k+j}]$ has a representative which can be taken to be a homotopy sphere is equivalent to the index being zero which in turn is equivalent to the statement α has a representative framed manifold sitting in the disk.

The above was for $k \equiv 0 \mod (4)$. For $k \equiv 1,2, \mod (8)$, the following sharper result can be deduced.

Theorem 3.7. *Let G^{k+j} be the non-trivial stable sphere bundle over S^k with $k \equiv 1, 2, \mod (8)$ $k \geq 5$. Then the best homology class realizable is equal to $2x$, where x is a generator of $H_k(G^{k+j}, Z)$. Moreover, the framed manifold may be taken to be a homotopy sphere.*

Proof. As in 3.5, one can derive a similar exact sequence with $a = J_{k-1}(1)$ and an integer d such that the image of $i^* = dZ$. From the exactness, $d =$ order $J_{k-1}(1)$ which has order 2. Hence, the best (non-zero) homology class realizable is $2x$, x a generator of $H_k(G^{k+j}, Z)$.

The rest follows from 2.1, since all the surgery obstructions are zero in these dimensions.

Remark. Consider a manifold M^{2k+j} of the type studied in Section 1 and 2. One may ask if an arbitrary framed manifold

(N^k, F^{k+j}) in M^{2k+j} can be written as a connected sum of framed manifolds (N^k, F_i^{k+j}) $i = 1,2,\cdots,s$ = rank of $H_k(M^{2k+j}, Z)$ such that N_i^k lies in the i-th handle and no other i.e., such that $KI(N_i^k, z_h^{k+j}) = 0$ except for $i = h$. The following example shows that there is a framed manifold which is not decomposable.

Take the $8+j$ dimensional disk D^{8+j}, $j \geq 4$, and attach two handles $D_i^4 \times D_i^{4+j}$ of index 4, $i = 1,2$, with associated sequence (8, 1). Neglecting the handle $D_2^4 \times D_2^{4+j}$ and looking at the attachment of the handle $D_1^4 \times D_1^{4+j}$ gives a disk bundle E_1^{4+j} over S^4. By 3.5, the best non-zero homology class in $H_4(E_1^{4+j}, Z) = Z$ given by a framed manifold not touching the boundary is $3x_1$, x_1 a generator of $H_4(E^{4+j}, Z)$. Similarly neglecting the handle $D_1^4 \times D_1^{4+j}$ and looking at only the handle $D_2^4 \times D_2^{4+j}$ gives a disk bundle E_2^{4+j} and the best non-zero homology class realizable by a framed manifold is $24x_2$, x_2 a generator of $H_4(E_2^{4+j}, Z)$.

By 1.7, one has the exact sequence:

$$0 \longrightarrow [M^{8+j}, S^{4+j}] \xrightarrow{i^*} Z \oplus Z \xrightarrow{a^*} Z_{24} \longrightarrow 0$$

where i^* is the map given by sending the framed manifold (N^4, F^{4+j}) to $[KI(N^4, z_1^{4+j}), KI(N^4, z_2^{4+j})]$ in $Z \oplus Z$, and $a = J_3(8) + J_3(1)$. One can show that the kernel of a^* is equal to the subgroup $(2, 8) \oplus (3, 0)$ where $(2, 8)$ and $(3, 0)$ also stand for the subgroups generated by these elements respectively. However, $(2, 8) \oplus (3, 0)$ contains the group $(3, 0) \oplus (0, 24)$ as a proper subgroup. Since $(2, 8)$ is in the image of i^*, there is a framed manifold (N^4, F^{4+j}) such that $KI(N^4, z_1^{4+j}) = 2$ and $KI(N^4, z_1^{4+j}) = 8$. But, $(2, 8)$ cannot be written as any linear combination of the elements $(3, 0)$ and $(0, 24)$, whence the manifold N^4 above is indecomposable.

It is interesting to note that N^4 above has index equal to 16.

Section 4. Manifolds with Six Critical Values. In this section, manifolds M^{2k+j} with the following data on its handle body decomposition will be examined. M^{2k+j} admits a Smale-Wallace function which has only critical values of index $0, k, k+1, k+j-1, k+j$, and $2k+j$ such that for each critical value of index $0,k$ there is only one corresponding critical point on the manifold M^{2k+j}.

The methods of the previous sections will be employed to show, knowledge of the cohomotopy group of a submanifold derived from the k skeleton of M^{2k+j}, forces very strong conditions on the way the $k+1$ skeleton is to be constructed, at least for $k \equiv 0 \mod (4)$.

Henceforth in this section, M^{2k+j} shall mean the manifold already described and $k = 4r > 0$ with j assumed larger than 4. By Section 3, C^k is represented by a $k+j$ disk bundle E^{k+j} over S^k and is classified by some element in $\pi_{k-1}(SO) = Z$ which will be denoted by m. Taking the double of C^k gives a $k+j$ sphere bundle G^{k+j} over S^k. C^{k+1} is obtained by attaching handles of index $k+1$ via embeddings h_i of $S^k \times D^{k+j-1}$ into the boundary bE^{k+j} of E^{k+j} (which is a $k+j-1$ sphere bundle over S^k).

<u>Theorem 4.1.</u> <u>If</u> $m \neq 0$, <u>then the attaching map</u> $h_i : S^k \times D^{k+j-1} \longrightarrow bE^{k+j}$ <u>for a handle of index</u> $k+1$ <u>has its image contained in a disk in</u> bE^{k+j}.

<u>Proof.</u> The mapping h_i gives a framing F^{k+j-1} of the image $\overline{S}^k$ of $S^k \times 0$ in bE^{k+j}. Therefore, $(\overline{S}^k, F^{k+j-1})$ is a representative of a cohomotopy class in $[bE^{k+j}, S^{k+j-1}]$. By 2.3 the index of any framed manifold (N^k, F^{k+j-1}) representing some cohomotopy class

in $[bE^{k+j}, S^{k+j-1}]$ is given by:

$$I(N^k) = a_r s_r (2r-1)!m\, KI\left(N^k, z^{k+j-1}\right)$$

where z^{k+j-1} is the homology class in $H_k\left(bE^{k+j}, Z\right)$ given by a fibre S^{k+j-1} in bE^{k+j}.

However, $I(\bar{S}^k)$ is zero which implies by the formula, $KI\left(\bar{S}^k, z^{k+j-1}\right) = 0$. Hence, the mapping h_i is such that $h_i(S^k \times 0)$ has algebraic intersection with the belt sphere $\left(0 \times S^{k+j-1}\right)$ in bC^k equal to zero. Therefore, h_i restricted to $S^k \times 0$ is homologically trivial which implies h_i is homotopically trivial. By the assumption on j and k, h_i may be extended to an embedding of the disk D^{k+1} into bC^k. This completes the proof of 4.1 .

<u>Remark</u>. One can see that the condition $k = 4r$ was necessary above in order to work with the index. It would be useful if one had an integer valued invariant for $k \equiv 1,2 \mod (8)$. The above can be argued in a similar way for $k \equiv 1,2 \mod (8)$ using Stiefel-Whitney class which give mod 2 invariants, the results are not as decisive as those in 4.1 .

If the number of critical points of index $k+1$ is one, then by 4.1 with $m \neq 0$, the manifold C^{k+1} admits the following form. There are maps

$$k_1 : S^{k-1} \times D^{k+j} \longrightarrow bD^{2k+j}, \quad h_1 : S^k \times D^{k+j-1} \longrightarrow bD^{2k+j}$$

such that k_1 and h_1 have disjoint images and C^{k+1} is obtained by attaching handles of index k and $k+1$ to the boundary of the disk D^{2k+j}. Consequently, the homology of M^{2k+j} is given by $H_i(M^{2k+j}, Z) = Z$ for $i = 0, 1, 1+1, k+j-1, k+j, 2k+j$ and zero otherwise. As in Section 3, the maps k_1 and h_1 give elements $m_1 = m$, m_2 in $\pi_{k-1}(SO)$, $\pi_k(SO)$ and elements $J_{k-1}(m_1)$, $J_k(m_2)$ representing homotopy classes in π^S_{k-1}, π^S_k respectively. The framing of the spheres in bD^{2k+j} given by the maps k_1 and h_1, give by Thom-Pontriagin construction a map $a : S^{2k+j-1} \longrightarrow S^{k+j} \vee S^{k+j-1}$.

By 1.1 for $p > k+2$, $[M^{2k+j}, S^p]$ is isomorphic to $[C^{k+1}/bC^{k+1}, S^p]$. Under suitable conditions put on k and j, one can prove 1.4, and therefore find a $2k+j-2$ manifold N^{2k+j-2} such that $S^2(N^{2k+j-2}/bN^{2k+j-2})$ is homeomorphic to C^{k+1}/bC^{k+1} so that $[C^{k+1}/bC^{k+1}, S^p]$ is an abelian group.

Theorem 3.2. Suppose $m_1 \neq 0$, $p > k+2$, $k = 4r > 0$. Then one has for $[M^{2k+j}, S^p]$:

(1) For $k \equiv 4 \mod (8)$,

$$[M^{2k+j}, S^p] = [G^{k+j}, S^p] \oplus [S^{k+j-1}, S^p].$$

(2) For $k \equiv 0 \mod (8)$, one has an exact sequence:

$$[S^{2k+j}, S^p] \xrightarrow{q^*} [M^{2k+j}, S^p] \xrightarrow{i^*}$$

$$\longrightarrow [S^{k+j} \vee S^{k+j-1}, S^p] \xrightarrow{a^*} [S^{2k+k-1}, S^p].$$

Proof.

Case 1. If $k \equiv 4 \mod (8)$, then $\pi_k(SO) = 0$, and therefore the $k+1$ handle is attached untwisted which gives a framed sphere (S^{k+1}, F^{k+j-1}) representing a generator of $H_{k+1}(M^{2k+j}, Z)$.

In C^{k+1}, there is the $k+j-1$ belt disk of the $k+1$ handle which is a sphere S^{k+j-1} in C^{k+1}/bC^{k+1}. Consider the cofibration,

$$S^{k+j-1} \xrightarrow{i} C^{k+1}/bC^{k+1} \xrightarrow{q} (C^{k+1}/bC^{k+1})/S^{k+k-1}.$$

Note that $(C^{k+1}/bC^{k+1})/S^{k+j-1}$ is homeomorphic to $C^k/bC^k = T(E^{k+j})$.

The above combined with the Puppe sequence and the half exact functor $[\ \ , S^p]$ gives the exact sequence of abelian groups.

$$[\Psi(E^{k+j}), S^p] \xrightarrow{q^*} [C^{k+1}/bC^{k+1}, S^p] \xrightarrow{i^*} [S^{k+j-1}, S^p].$$

Using the framed manifold (S^k, F^{k+j-1}), define a homomorphism r^* as in the proof of 3.2 from $[S^{k+j-1}, S^p \longrightarrow [C^{k+1}/bC^{k+1}, S^p]$ which splits the above sequence. For $p > k+2$, one has $[C^{k+1}/bC^{k+1}, S^p] = [M^{2k+j}, S^p]$ and $[\Psi(E^{k+j}), S^p]$ is isomorphic to $[G^{k+j}, S^p]$, where G^{k+j} has already been described. Along with the splitting this gives Case 1.

<u>Case 2</u>. Let X be the mapping cylinder of the map a. In a similar way as in 1.6, X is homeomorphic to C^{k+1}/bC^{k+1} which allows one to replace X by C^{k+1}/bC^{k+1} in the mapping cylinder sequence of X to get:

$$S^{2k+j-1} \xrightarrow{a} S^{k+j} \vee S^{k+j-1} \xrightarrow{i} C^{k+1}/bC^{k+1} \xrightarrow{q} S^{2k+j} \xrightarrow{Sa} \cdots$$

Applying now $[\ \ , S^p]$ to the first four terms of this sequence along with the isomorphism of $[C^{k+1}/bC^{k+1}, S^p]$ with $[M^{2k+j}, S^p]$ valid for $p > k+2$ gives Case 2.

Remark. For the case $k \equiv 4 \mod (8)$, (1) above gives $[M^{2k+j}, S^p]$ in terms of the maps of spheres bundles and spheres into spheres. If one employs the description of maps of sphere bundles into spheres given in Section 3, an extensive determination of $[M^{2k+j}, S^p]$ can be given.

Appendix. R^n will denote the topological, peicewise linear, or smooth Euclidean n space, in the appropriate category.

Consider $PL_n(R)$, $\overline{PL}_n(R)$, as given on pages 19, 22, of [14]. Represent by $Top_n(R)$ and $\overline{Top}_n(R)$ the corresponding topological analogues. Let $f \in \pi_{k-1}(PL_n(R))$ and $\bar{f} \in \pi_{k-1}(\overline{PL}_n(R))$. Using the maps f and $\bar{f}$, and the construction given at the beginning of Section 3 for smooth disk bundles, one can construct bundles which will be called the $PL_n - R^n$ - bundle and $\overline{PL} - R^n$ -block bundle over S^k respectively. Similarly statements hold for $Top_n(R)$ and $\overline{Top}_n(R)$. The various structural groups are related by the following commutative diagram:

$$\begin{array}{ccc} PL_n(R) & \xrightarrow{p\bar{p}} & \overline{PL}_n(R) \\ {\scriptstyle pt}\downarrow & & \downarrow{\scriptstyle \overline{pt}} \\ Top_n(R) & \xrightarrow{t\bar{t}} & \overline{Top}_n(R) \end{array}$$

where the maps arise by various inclusions (see [14]).

In 1.6, a description of the Hopf - Whitehead homomorphism was given. The same method works if one uses elements in $\pi_{k-1}(\overline{PL}_n(R))$ or $\pi_{k-1}(\overline{Top}_n(R))$. From this, one gets homomorphisms,

$$J_{k-1}^{\overline{Top}} : \pi_{k-1}(\overline{Top}_n(R)) \longrightarrow \pi_{k+n-1}(S^n); \quad J_{k-1}^{\overline{PL}} : \pi_{k-1}(\overline{PL}_n(R)) \longrightarrow \pi_{k+n-1}(S^n)$$

respectively such that the following diagram is commutative.

$$\begin{array}{ccc} \pi_{k-1}(\overline{PL}_n(R)) & \xrightarrow{\overline{pt}_*} & \pi_{k-1}(\overline{Top}_n(R)) \\ & {\scriptstyle J_{k-1}^{\overline{PL}}} \searrow \quad \swarrow {\scriptstyle J_{k-1}^{\overline{Top}}} & \\ & \pi_{k+n-1}(S^n) & \end{array}$$

$\overline{pt}_*$ is the map induced from the forgetful map $\overline{pt} : \overline{PL}_n(R) \longrightarrow \overline{Top}_n(R)$.

Now $J_{k-1}^{\overline{PL}}$ is onto apriori $J_{k-1}^{\overline{Top}}$ if $k-1$ is not of the form $2^m - 2$ for some integer $m > 0$, $n > k$. Indeed, by 2.1 each element in π_{k-1}^S for $k-1 \neq 2^m - 2$, is represented by a framed smooth homotopy sphere. By the PL version of Poincare conjecture, the result follows.

Remark. In [14], page 12, for n dimensional disk block bundles, an object is defined which is denoted here by $\overline{PL}_n(D)$. Moreover, a map $i_*: \overline{PL}_n(D) \longrightarrow \overline{PL}_n(R)$ is constructed which is a homotopy equivalence. Hence, $J^{PL}_{k-1} \circ i_* : \pi_{k-1}(\overline{PL}_n(D)) \longrightarrow \pi^S_{k-1}$ is onto as well for $k-1 \neq 2^m - 2$ $(m > 0)$, and for brevity $J^{PL}_{k-1} \circ i_*$ was denoted by J^{PL}_{k-1} in Proposition 1.8 .

For $k = 4r > 0$, a definition of the r-th Pontriagin class of a $Top_n - R^n$ - bundle will now be given.

H^n will represent a topological block R^n bundle over S^k, $n > k$ classified by the element $\alpha \in \pi_{k-1}(\overline{Top}_n(R))$.

Consider the collection of all pairs (N^k, i) where N^k is a closed compact oriented topological manifold and i is an embedding of $N^k \times R^n$ into H^n. Define a map $g*$ from the collection of such pairs to the integers Z by $g*(N^k, i) = \text{index}\ (N^k)$. If $-N^k$ denotes N^k with its orientation reversed, then $g*(-N^k, i) = -g*(N^k, i)$. Also, 0 belongs to the image of $g*$, because one has an embedding of the sphere S^k into a disk in H^n. These remarks combine to prove image of $g*$ is a group and therefore, image $g^* = dZ$ for some non-negative integer d.

Using the Cairns - Hirsch Theorem and the proof of 2.6, it follows that for each (N^k, i) N^k is an almost smoothable, almost parallelizable manifold, $k > 4$.

Definition 4.1. The rational Pontriagin class of a $\overline{\mathrm{Top}}_n - R^n$ - bundle H^n over S^k, $k = 4r > 0$, classified by $\alpha \in \pi_{k-1}(\overline{\mathrm{Top}_n}(R))$, $n > k$, is defined by:

$$P(H^n) = 1 + p_r = 1 + [(d(2r!)/(2^{2r}(2^{2r-1} - 1)B_r o(a)))] \circ X$$

where 1 is a generator of $H^0(S^k, Z)$, X an oriented generator of $H^k(S^k, Z)$, B_r the r-th Bernoulli number, $a = J_{k-1}^{\overline{\mathrm{Top}}}(\alpha)$, and $o(a) = (J_{k-1}^{\overline{\mathrm{Top}}}(\alpha))$ is the order of $J_{k-1}^{\overline{\mathrm{Top}}}(\alpha)$ in π_{k-1}^S.

As one sees from the above equation, the fact that the coefficient of X is rational and not an integer is reflected in part that one must divide by the order of the element $J_{k-1}^{\overline{\mathrm{Top}}}(\alpha)$.

If H^n is smooth vector bundle i.e., if α is in the image of $st_*: \pi_{k-1}(SO_n) \longrightarrow \pi_{k-1}(\overline{\mathrm{Top}_n}(R))$, it will now be shown that the above definition agrees with the ordinary Pontriagin class.

<u>Proposition 4.2.</u> <u>If</u> H^n <u>is a smooth vector bundle, the previous definition gives the smooth Pontriagin class.</u>

<u>Proof.</u> If H^n is smooth, then by 2.5

$$P_r(H^n) = a_r(2r-1)!\, m \circ X$$

where $m \in \pi_{k-1}(SO) = Z$ is the characteristic class and X is as above. By Corollary 3.6, the number d for smooth H^n is equal to

$$a_r m(\overline{m}, B_r) 2^{2r-1}(2^{2r-1} - 1)(B_r/r).$$

From the proof of 3.5, order of $J_{k-1}(m) = (\overline{m}, B_r)$. By a short computation, it is easily verified that the two definitions agree.

REFERENCES

1. Adams, J. R., On the groups J(X) III. Topology 3 (1966), 193-222.

2. Barratt, M. G., Mahowald, M. E., The metastable homotopy of O(m). Bull. Amer. Math. Soc. 70 (1964), 758-760.

3. Browder, W., The Kervaire invariant of framed manifolds and its generalization. Ann. Math. 90 (2) (1969), 157-186.

4. Haefliger, A., Plongements differentiables de variétés dan variétés. Comment. Math. Helv. 36 (1961), 47-82.

5. ____________, Knotted 4k - 1 spheres in 6k space. Ann. of Math. (2) 75 (1962), 452-466.

6. Hirzebruch, F., Neue topologische methoden in der algebraischen geometrie. Springer-Verlag, Berlin, 1962.

7. Hudson, J. F. P., Piecewise linear topology. W. A. Benjamin, Inc. New York, 1969.

8. Kervaire, M., A note on obstructions and characteristic classes. Amer. J. Math. 81 (1959), 773-784.

9. ____________, An interpretation of G. Whitehead's generalization of Hopf's invariant. Ann. of Math. (1) 69 (1959), 345-365.

10. Kervaire, M., Milnor, J., Groups of homotopy spheres I., Ann. of Math. (2) 77 (1963), 504-532.

11. Kosinski, A., On the inertia group of π - manifolds. Amer. J. Math. 89 (2) (1967), 227-248.

12. Levine, J., A classification of differentiable knots. Ann. of Math. (2) 82 (1965), 15-50.

13. Quillen, D., The Adams conjecture. Topology 10 (1971), 67-80.

14. Rourke, C. P., Sanderson, B. J., Block bundles I. Ann. of Math. (2) 87, (1968), 1-28.

15. Serre, J., Groupes d'homotopie et classes de groupes abéliens. Ann. of Math. (2) 58 (1953), 258-294.

A SURVEY OF DIFFEOMORPHISM GROUPS

by

Edward C. Turner

§0.) Introduction

This article is intended to serve as an introduction to that part of the study of diffeomorphism groups that relates to: a) their structure as topological spaces, including homotopy type, and b) the interest in and calculation of the group of path components = isotopy classes. (I'll not discuss results in dynamical systems since the techniques are more analytic and I'm not familiar with the area.) The bibliography is fairly complete and includes references to the related studies of groups of homotopy equivalences, homeomorhisms and PL homeomorphisms. The notation will be as follows: $\mathcal{D}(M)$ and $\mathcal{D}(M,N)$ are the groups of diffeomorphisms of M (resp. that are the identity on N) with the uniform C^r topology for some $r \geq 2$; $\mathcal{D}_0(M)$ is the component of the identity; $\mathcal{D}(M) = \mathcal{D}(M)/\mathcal{D}_0(M) = \pi_0(\mathcal{D}(M))$; $\mathcal{D}^{\pi}(M)$ is the subgroup of $\mathcal{D}(M)$ of classes with null-homotopic representatives; $f \sim g$ means that f and g are isotopic; i.e., there exists a level preserving diffeomorphism $H: M \times I \longrightarrow M \times I$ such that $H(m,0) = (f(m),0)$ and $H(m,1) = (g(m),1)$; $f \sim_p g$ means that f and g are pseudo-isotopic = quasi-isotopic = concordant; i.e., there exists an H as above which may not be level preserving. All manifolds are assumed to be compact and oriented and diffeomorphisms orientation preserving.

The subject of §1 is the relation between pseudo-isotopy classification of diffeomorphisms and diffeomorphism classification of manifolds. In §2, the Hilbert (or Frechet) manifold structure of $\mathcal{D}(M)$ is described along with very interesting recent results on its homotopy type. §3 describes present state of the pseudo-isotopy/isotopy question. §4 deals with the subgroups $\mathcal{D}^{\pi}(M)$. §5 briefly indicates the results of [A1] on the "concordance homotopy groups" of $\mathcal{D}(M)$. Finally, §6 includes a summary of results I find interesting but which do not fit in one of the above classifications. A number of open questions are sprinkled through the exposition.

§1.) Classification problems

It is well known [K5] that $\Theta_n \cong \mathcal{D}(S^{n-1})(n \neq 3,4)$ and this was the first such calculation. It depends on the very special fact that the set of homotopy spheres has a natural group structure. Although the assignment $d \longrightarrow W(d) = W \cup_d W$ makes sense in general for $d \in \mathcal{D}(M)(M = \partial W)$, the set of manifolds so obtained usually cannot be given even a semi-group structure. However, one can use this construction to distinguish diffeomorphisms up to pseudo-isotopy since it is easy to show that $W(d_1) \not\approx W(d_2)$ implies that $d_1 \not\simeq_p d_2$. This approach has been successfully employed to classify manifolds with 2 non-vanishing homology groups (other than 0 and n) [T3,W3,S1]. It would be interesting to know if this is theoretically sufficient.

Question: If M is a boundary and $d_1 \not\simeq_p d_2$ in $\mathcal{D}(M)$, does there exist a W, $M = \partial W$, such that $W(d_1) \not\approx W(d_2)$?

A related construction converting questions about pseudo-isotopy of diffeomorphisms into questions about diffeomorphism of manifolds is the mapping torus construction. If $d \in \mathcal{D}(M)$, let $T(d) = M \times I/(m,0) \sim (d(m),1)$. Clearly, if $d_1 \sim_p d_2$ then $T(d_1) \cong T(d_2)$. Here there is a partial converse [B3]; if M is simply connected and $D: T(d_1) \to T(d_2)$ is a diffeomorphism such that p_1 is homotopic to $p_2 \circ D$, $\{p_i: T(d_i) \to S^1$ the natural map$\}$, then d_1 and d_2 are pseudo-isotopic. These techniques can often be applied to give special results and even to get complete calculations in special cases, but fall short of dealing completely with diffeomorphism groups because of the loss of the group structure.

§2.) $\mathcal{D}(M)$ as a topological space

$\mathcal{D}(M)$ is a Hilbert or Fréchet manifold. The local model is $\Gamma(\tau(m))$, sections of the tangent bundle of M, as follows [P1]: let $\psi: \Gamma(\tau(m)) \to C^\infty(M,M)$ by integration; i.e., $\psi(s)(m) = \gamma(1)$ where γ is an integral curve of s through m. For s sufficiently small with respect to some Riemannian metric, $\psi(s)$ is a diffeomorphism, essentially by the inverse function theorem. Since the space $\Gamma_\varepsilon(\tau(M))$ of small sections is isomorphic to $\Gamma(\tau(M))$, $\psi|\Gamma_\varepsilon(\tau(M))$ is the asserted local model. For M compact and $r < \infty$, $\Gamma(\tau(M))$ is a Hilbert space with $\langle s,s' \rangle = \int_M \langle s_r(m), s_r'(m) \rangle d\mu$, where s_r is the r-jet associated with s and $\langle \, , \, \rangle$ is some Riemannian metric for the bundle of r-jets. If $r = \infty$, the sequence of associated norms defines a Frechet structure. The transition maps can be checked to be almost as smooth as the sections (C^r sections give C^{r-1} transition maps) so that $\mathcal{D}(M)$ has the structure of an infinite dimensional smooth manifold [L3]. (This smooth structure has not entered into the results described in this article.) It follows from the Hilbert manifold structure that:

a) $\mathcal{D}(M)$ has the homotopy type of a countable CW complex [P2],

b) its homeomorphism type is determined by its homotopy type [H4]. {It's not hard to show that its homotopy type is independent of r, $2 \leq r \leq \infty$}. A natural question is whether it has finite type. The answer is very different in low and high dimensions. In dimension 2, $SO(3) \subset \mathcal{D}_0(S^2)$, $T^2 \subset \mathcal{D}_0(T^2)$ (by group action) and $* \subset \mathcal{D}_0(M)$, (genus $M > 2$) are deformation retracts [E1,H1,S5,S11: non-oriented case in B1,B2,C11]. Partial negative results have been obtained in higher dimensions in [A1]. Consider $\mathcal{D}(S^n)$; as it is homotopy equivalent to $SO_{n+1} \times \mathcal{D}(D^n,S^{n-1})$, it has finite type if and only if $\mathcal{D}(D^n,S^{n-1})$ does. They show that <u>for $n > 7$, $\mathcal{D}(D^n,S^{n-1})$ does not have finite type</u> (in fact it is not even dominated by a finite CW complex). The method of proof is easily described (but not so easily executed!) $\mathcal{D}(D^n,S^{n-1})$ is homotopy commutative since any pair of diffeomorphisms can be supported (canonically) on disjoint discs. Thus by Hubbeck's Theorem [H11], if $\mathcal{D}(D^n,S^{n-1})$ has finite type, it is homotopy equivalent to the n-torus $S^1 \times \cdots \times S^1$. The proof, then, involves detecting non-trivial elements of $\pi_i(\mathcal{D}(D^n, S^{n-1}))$ for $i > 1$ or non-trivial torsion elements in $\pi_1(\mathcal{D}(D^n, S^{n-1}))$. Certain more general information can be obtained by considering the inclusion $\mathcal{D}(D^n, S^{n-1}) \longrightarrow \mathcal{D}(M)$ (extend as the identity outside a disc), but the general question is still open.

<u>Question: Is there a manifold M^n ($n \geq 5$, say) for which $\mathcal{D}(M)$ has finite type?</u>

<u>Remarks</u>:

i) Lawson [L2] has a short proof that if $1 \leq j \leq i-2$, i and j both odd and M a j-dimensional manifold, then $\pi_{i-j}(\mathcal{D}(M \times S^i))$ has elements of infinite order, and in fact, they are of infinite order when considered in the PL and TOP categories (not true of those detected in [A1].

ii) If the pointwise C^k topology is used, $SO(n+1)$ is a deformation retract of $\mathcal{D}(S^n)$ [R1]. However, this topology is not significant for the purposes of differential topology.

From a different angle, there is the result of Morlet [M2], later generalized by Burghelea and Lashof [B6], that $\mathcal{D}(D^n, S^{n-1})$ has the homotopy type of $\Omega^{n+1}(PL_n/O_n)$! Burghelea and Lashof's Theorem can be roughly phrased as follows: let B be the result of replacing the fiber of $\tau(M)$ with PL_n/O_n and Γ_N be the sections of B which are trivial on N. Then there exists a map i

$$i:\{PL(M,N)/\mathcal{D}(M,N)\} \longrightarrow \Gamma_N$$

which is a homotopy equivalence onto some set of components of Γ_N. This is proven by converting from (PL and smooth) automorphisms of M to the appropriate bundle maps and then classifying them. (Of course, all this must be done semi-simplicially so that $PL(M,N)$ makes sense.) Now, $PL(D^n, S^{n-1})$ is contractible by the Alexander trick and B is trivial, so we have:

$$B_{\mathcal{D}(D^n,S^{n-1})} \text{ (the classifying space)} \sim PL(D^n, S^{n-1})/\mathcal{D}(D^n, S^{n-1}) \xrightarrow[i]{} \Omega^n(PL_n/O_n)$$

and by looping

$$\mathcal{D}(D^n,\ S^{n-1}) \sim \Omega^{n+1}(PL_n/O_n)\ .$$

<u>Question</u>: <u>Under what circumstances is Γ trivial</u>?

3.) <u>Pseudo-isotopy/isotopy</u>

The first and main theorem in this connection is Cerf's [C5]: <u>if M^n is simply connected and $n \geq 5$, then two diffeomorphisms of M are isotopic if and only if they are pseudo-isotopic</u>. (This, by the way, is a necessary part of the isomorphism $\Theta_n \cong \pi_0(\mathcal{D}(S^{n-1}))$.) The non-simply connected case is more complicated. Let $\mathcal{P}(M \times I) = \mathcal{D}(M \times I,\ M \times 0)$ be the set of pseudo-isotopies of M and

$\mathcal{P}(M \times I) \xrightarrow{e} \mathcal{D}(M)$ be restriction to $M \times 1$, so $e(\pi_0(\mathcal{P}))$ is the group of diffeomorphisms pseudo-isotopic to the identity. Wagoner [W1] and Hatcher [H2,H3] have shown that <u>$\pi_0(\mathcal{P})$ is an abelian group which depends only on $\pi_1(M)$ and the action of $\pi_1(M)$ on $\pi_2(M)$</u>. They have given an algebraic description of this group and it is known to be non-zero in some cases; e.g., if $M = W \times S^1$, $Wh(\pi_1(W)) \neq 0$ [S8]. The problem is therefore "reduced" to understanding the algebraic description, given in terms of algebraic K - theory.

§4.) <u>The subgroup $\mathcal{D}^{\pi}(M)$</u>

If $\pi_1(M) = 0$ and $\partial M = \phi$ (assumed only for simple exposition) consider $h\mathcal{D}(M) = \{h: M \times I \longrightarrow M \times I \mid h$ is a relative homotopy equivalence, $h|M \times 0 = id$, $h|M \times 1$ is a self diffeomorphism$\}$. Then the obvious map $\pi_0(h\mathcal{D}(M)) \longrightarrow h\mathcal{S}[M \times I, \partial(M \times I)]$ can be checked to be an isomorphism (using the h - cobordism theorem and pseudo-isotopy implies isotopy), so that the theory of surgery can be applied. In particular, there is an exact sequence

$$\begin{array}{ccccccccc} 0 & \longrightarrow & bP_{n+2} & \longrightarrow & h\mathcal{S}[M \times I, \partial(M \times I)] & \xrightarrow{\alpha} & [\Sigma M, G/O] & \longrightarrow & P_{n+1} \\ & & & & \downarrow \psi & & & & \\ & & & & \mathcal{D}^{\pi}(M) & & & & \end{array}$$

where ψ is restriction to $M \times 1$ and α is well known to be a homomorphism. It is now clear that $\mathcal{D}^{\pi}(M)$ is i) finitely generated ii) abelian if n is odd (so $bP_{n+2} = 0$) and iii) nilpotent if n is even (because $\psi(bP_{n+2}) \subset$ center of $\mathcal{D}^{\pi}(M)$). (Under appropriate reformulation, $\mathcal{D}^{\pi}(M)$ is nilpotent for $\pi_1(M) \neq 0$ as well [T4].)

Furthermore, $\mathcal{D}^{\pi}(M)$ can be calculated "mod Q" as follows: $[\Sigma M, G/O] \otimes Q \cong [\Sigma M, BO] \otimes Q \cong H^{4*}(\Sigma M, Q)$ together with Wang's observation [W5] that $\ker \psi$ is finite implies that rank $(\mathcal{D}^{\pi}(M) \otimes Q) =$ rank $(H^{4*}(\Sigma M, Q)) - t$, $t = 1$ if $(n+1) \equiv 0 \pmod 4$, $t = 0$ otherwise. Another approach to $\mathcal{D}^{\pi}(M)$ is described in [T2].

§5.) The concordance homotopy groups [A1]

In studying the higher homotopy groups of $\mathcal{D}(M)$ one is confronted with the following problem: given a level preserving diffeomorphism of $S^i \times M$, when does it extend to a level preserving diffeomorphism of $D^{i+1} \times M$? The usual techniques of differential topology apply much more naturally when the restriction of level preservation is not present. This observation motivates the definition given in [A1]: the ith concordance homotopy group $\pi_i(\mathrm{Diff};M)$ of $\mathcal{D}(M)$ is the set of diffeomorphisms of $S^i \times M$ (identity on $D^i \times M$) modulo those that extend over $D^{i+1} \times M$ (identity on "half of D^{i+1}" $\times M$) with a naturally defined group structure. Another way of thinking of it is as the ith homotopy group of the semi-simplicial complex whose k-simplices are diffeomorphisms of $\Delta^k \times M$ which are face preserving but may not preserve the first coordinate. So the zero-th concordance homotopy group is $\mathcal{D}(M)$ modulo the relation of pseudo-isotopy (= concordance). They define analogous objects in the PL, TOP and H (homotopy equivalence) categories and relative versions and derive long exact sequences like:

$$\cdots \longrightarrow \pi_i(\mathrm{Diff}:M) \longrightarrow \pi_i(H;M) \longrightarrow \pi_i(H,\mathrm{Diff};M) \longrightarrow \pi_{i-1}(\mathrm{Diff}:M) \longrightarrow \cdots$$

The interest here is the calculation of the relative term:

$\pi_i(H,\mathrm{Diff};M) \cong h\mathcal{S}[M \times D^i, M \times S^{i-1}]$ which fits in a familiar surgery exact sequence. A somewhat surprising corollary can be drawn from this: using the fact that the other terms in the surgery exact sequence depend only on the homotopy type of M, $\pi_i(\mathrm{Diff}:M)$ depends, up to extension, only on the homotopy type of M.

Question: How much of $\mathcal{D}(M)$ depends only on the homotopy type of M?

Another natural question is how exactly are the concordance homotopy groups related to the ordinary ones. This is immensely difficult - it is a higher dimensional version of the pseudo-isotopy/isotopy problem.

§6.) Other results

Several authors [K4,L5,S2,T3] computed $\mathcal{D}(S^p \times S^q)$ and PL and H analogues about the same time: if $q < p$, then

$$\mathcal{D}(S^p \times S^q) = (FC_p^{\,q+1} \oplus \Theta_{p+q+1}) \times_\phi FC_q^{\,p+1}$$

(a semi-direct product) where FC_i^j is the group of framed S^i's in S^{j+i}. Hodgsen [H6,H8,H9] has described a calculation of $\mathcal{D}(M)$ in the case that M is a "thickening" of a complex K^k,

$k \leq n-3$, K^k a $(2k-n+2)$-connected suspension: e.g., M a sufficiently stable disc bundle over a suspension. Wells [W6] has shown how to use this to describe $\mathcal{D}(\partial M)$, ∂M as above.

In a different vein, Epstein [E3] and Thurston [unpublished] have proven that $\mathcal{D}_0(M)$ is a simple group: Epstein - $[\mathcal{D}_0(M), \mathcal{D}_0(M)]$ is simple and Thurston - $\mathcal{D}_0(M) = [\mathcal{D}_0(M), \mathcal{D}_0(M)]$. And Mather [M1] showed that if G is the group of compactly supported diffeomorphisms of $\mathbb{R}^n$, $H^r(G, \mathbb{Z}) = 0$ for $r > 1$, where we mean the group cohomology of the discrete group G.

Finally, I would like to state a result which is likely to be a useful tool in understanding $\mathcal{D}^{\pi}(M)$ [T2].

<u>Theorem</u>: Suppose $n \geq 5$, M^n is 2 connected and $N \subset M$. If f and g are homotopic diffeomorphisms of M^n, then $f|N$ and $g|N$ are isotopic modulo local knots in the sense that they differ by an embedding of D^n in D^m.

BIBLIOGRAPHY

[A1] P. L. Antonelli, D. Burghelea, P. J. Kahn, The concordance homotopy groups of geometric automorphism groups, Springer lecture notes #215.

[A2] ______________, Gromoll groups, Diff (S^n) and bilinear constructions of exotic spheres, Bull. Am. Math. Soc. 76 (1970), 722-727.

[A3] ______________, The non-finite homotopy type of some diffeomorphism groups, Topology 11 (1972), 1-49.

[A4] M. K. Agoston, On handle decompositions and diffeomorphisms, Trans. Am. Math. Soc. 137 (1969), 21-36.

[A5] M. Arkowitz, C. Curjel, The group of homotopy equivalences of a space, Bull. Am. Math. Soc. 70 (1964), 293-296.

[A6] A. Asada, Contraction of the group of diffeomorphisms of $\mathbb{R}^n$, Proc. Japan Acad. 41 (1965), 273-276.

[B1] J. S. Birman, D. R. J. Chillingsworth, On the homeotopy group of a non-orientable surface, Proc. Camb. Philos. Soc. 71 (1972), 437-448.

[B2] J. S. Birman, H. Hilden, Isotopies of homeomorphisms of Riemann surfaces and a theorem about Artin's braid group, Bull. Am. Math. Soc. (6) 78 (1972), 1002-1004.

[B3] W. Browder, Diffeomorphisms of 1 - connected manifolds, Trans. Am. Math. Soc. 128 (1967), 155-163.

[B4] W. Browder, T. Petrie, Diffeomorphisms of manifolds and semi-free actions on homotopy spheres, Bull. Am. Math. Soc. 77 (1971), 160-163.

[B5] M. Brown, Constructing isotopies in non-compact 3 - manifolds, Bull. Am. Math. Soc. (3) 78 (1972), 461-464.

[B6] D. Burghelea, R. Lashoff, The homotopy type of the space of diffeomorphisms, Part I, preprint.

[C1] J. Cerf, Groupes d'automorphismes et groupes de difféomorphismes des variétés compactes de dimension 3. Bull. Soc. Math. France 87 (1959), 319-329.

[C2] ______, Groupes d'homotopie locaux et groupes d'homotopie mixtes des espaces bitopologiques. Presque n - locale connexion. Applications. C. R. Acad. Sci. Paris 253 (1961), 363-365.

[C3] ______, Isotopie et pseudo-isotopie, Proc. Int. Cong. Math. Moscow (1966), 429-437.

[C4] ______, La nullité de $\pi_0(\mathrm{Diff}\ S^3)$. Theorems de fibration des espaces de plongements. Applications. Séminaire Henri Cartan, 1962/63, Exp. 8, 13 pp. Secrétariat mathématiques, Paris, 1964:
1) Position du problem, Exp. 9-10, 27 pp.:
2) Espaces fonctionnels liés aux decomposition d'une sphere plongée dans $\mathbb{R}^3$ Exp. 20, 29 pp.
3) Construction d'une section pour le revêtement . Exp. 21, 25 pp.

[C5] ______, The pseudo-isotopy theorem for simply connected differentiable manifolds, Manifolds - Amsterdam 1970 Springer lecture notes #197, p. 76-82.

[C6] ______, La stratification naturelle des espaces de fonctions différentiables réeles et le théorèm de la pseudo-isotopie. Haut Études Sci. Publ. Math. No. 39 (1970) 5-173.

[C7] ______, Topologie de certains espaces de plongements, Bull. Soc. Math. France 89 (1971), 227-380.

[C8] A. Chenciner, Pseudo-isotopies différentiables and pseudo-isotopies linéaires par morceaux. C. R. Acad. Sci. Paris Sér. A - B 270 (1970), A1312 - A1315.

[C9] A. Chenciner, F. Laudenbach, Contribution à une théorie de Smale à un paramètre dan le cas non-simplement connexe, Ann. Sci. École Norm. Sup. (4) 3 (1970), 109-478.

[C10] J. A. Childress, Restricting isotopies of spheres, Pac. J. of Math. (2) 45 (1973).

[C11] D. R. J. Chillingsworth, A finite set of generators for the homeotopy groups of a non-orientable surface, Proc. Camb. Philos. Soc. 65 (1969), 409-430.

[E1] C. J. Earl, J. Eells, The diffeomorphism group of a compact Riemann Surface, Bull. Am. Math. Soc. 73 (1967) 557-559.

[E2] D. G. Ebin, J. Marsden, Groups of diffeomorphisms and the motion of an incompressible fluid. Ann. of Math. (2) 92 (1970), 102-163.

[E3] D. B. A. Epstein, The simplicity of certain groups of homeomorphisms, Compositio Math. 22 (1970), 165-173.

[G1] R. Geoghegan, Manifolds of piecewise linear maps and a related normed linear space, Bull. Am. Math. Soc. 77 (1971), 629-632.

[G2] __________, On spaces of homeomorphisms embeddings and functions I, Topology 11 (1972), 159-177.

[G3] H. Gluck, Embeddings and automorphisms of open manifolds, Topology of Manifolds (Proc. Inst. Univ. Georgia, Athens, Georgia 1969), 394-406.

[G4] ______, Restriction of isotopies, Bull. Am. Math. Soc. 69 (1963), 78-82.

[H1] M. E. Hamstrom, Homotopy groups of the space of homeomorphisms of a 2-manifold, Illinois J. of Math. 10 (1966), 563-573.

[H2] A. Hatcher, A K_2 obstruction for pseudo-isotopies, Ph.D. thesis, Stanford 1971.

[H3] ______, The second obstruction for pseudo-isotopies, Bull. Am. Math. Soc. (6) 78 (1972), 1005-1008.

[H4] D. W. Henderson, R. Schori, Topological classification of infinite dimensional manifolds by homotopy type, Bull. Amer. Math. Soc. (1) 76 (1970), 121-124.

[H5] J. P. E. Hodgson, Automorphisms of meta-stably connected PL-manifolds, Proc. Camb. Philos. Soc. 69 (1971), 75-77.

[H6] ______, Automorphisms of thickenings, Bull. Am. Math. Soc. 73 (1967), 678-681.

[H7] ______, A generalization of "Concordance of PL homeomorphisms of $S^p \times S^q$", Can. J. Math. (3) 24 (1972), 426-431.

[H8] ______, Poincaré complex thickenings and concordance obstructions, Bull. Am. Math. Soc. 76 (1970), 1039-1043.

[H9] ______, Obstructions to concordance for thickenings, Invent. Math. 5 (1968), 292-316.

[H10] W. C. Hsiang, W. Y. Hsiang, On compact subgroups of the diffeomorphism groups of Kervaire spheres, Ann. of Math. (2) 85 (1967), 359-368.

[H11] J. R. Hubbeck, On homotopy commutative H-spaces, Topology 8 (1969), 119-126.

[H12] J. F. P. Hudson, Concordance and isotopy of PL embeddings, Bull. Am. Math. Soc. 72 (1966), 534-535.

[H13] ______________, Concordance, isotopy and diffeotopy, Ann. of Math. (2) 91 (1970), 425-448.

[H14] ______________, Piecewise linear embeddings and isotopies, Bull. Ann. Math. Soc. 72 (1966), 536-537.

[H15] J. F. P. Hudson, W. B. R. Lickorish, Extending piecewise linear concordances, Quart. J. of Math., Oxford Ser (2) 22 (1971), 1-12.

[H16] L. S. Husch, Homotopy groups of PL embedding spaces, Pac. J. Math. 33 (1970), 149-155.

[H17] __________, Local algebraic invariants for Δ-sets, Rocky Mountain J. Math. (2) 2 (1972), 289-298.

[H18] L. S. Husch, T. B. Rushing, Restriction of isotopies and concordances, Mich. Math. J. 16 (1969), 303-307.

[K1] D. W. Kahn, A note on H-equivalences, Pac. J. of Math. (1) 42 (1972), 77-80.

[K2] __________, The group of stable self equivalences, Topology 11 (1972), 133-140.

[K3] P. J. Kahn, Self equivalences of (n-1) connected 2n manifolds, Math. Ann. 180 (1969), 26-47.

[K4] M. Kato, A concordance classification of PL homeomorphisms of $S^p \times S^q$, Topology 8 (1969), 371-383.

[K5] M. Kervaire, J. Milnor, Groups of homotopy spheres I, Ann. of Math. 2 (1963), 504-537.

[L1] W. A. LaBach, On diffeomorphisms of the n - disc, Proc. Japan Acad. 43 (1967), 448–450.

[L2] T. C. Lawson, Some examples of non-finite diffeomorphism groups, Proc. Am. Math. Soc. (2) 34 (1972), 570-572.

[L3] J. A. Leslie, On a differential structure for the group of diffeomorphisms, Topology 6 (1967), 263-271.

[L4] J. Levine, Inertia groups of manifolds and diffeomorphisms of spheres, Amer. J. Math. 92 (1970), 243-258.

[L5] ______, Self equivalences of $S^n \times S^k$, Trans. Am. Math. Soc. 143 (1969), 523-543.

[M1] J. Mather, The vanishing of the homology of certain groups of homeomorphisms, Topology 10 (1971), 297-298.

[M2] C. Morlet, Lissage des homéomorphisms, CR Acad. Sci. Paris Sér. A - B 268 (1969), A1323-1326.

[N1] S. P. Novikov, Differentiable sphere bundles (Russian) Izv. Akad. Nauk SSSR Ser. Mat. 29 (1965), 71-96.

[N2] ______, Homotopy properties of the group of diffeomorphisms of the sphere (Russian), Dokl. Akad. Nauk. SSSR 148 (1963), 32-35: (English) Doklady 4 (part 1) (1963), 27-31.

[O1] P. Olum, Self equivalences of pseudo-projective planes II. Simple equivalences. Topology 10 (1971), 257-260.

[O2] H. Omori, Local structures on groups of diffeomorphisms, J. Math. Soc. Japan 24 (1972), 60-88.

[O3] ______, On the group of diffeomorphisms of a compact manifold, Global Analysis (Proc. Sympos. Pure Math. Vol. XV, Berkeley 1968), AMS 1970.

[P1] R. Palais, Local triviality of the restriction map for embeddings, Comment. Math. Helv. 34 (1960), 305-312.

[P2] ________, Homotopy theory of infinite dimensional manifolds, Topology 5 (1966), 1-16.

[R1] S. Robertson, Retracting diffeomorphisms of spheres, Proc. Am. Math. Soc. 24 (1970), 57-59.

[S1] H. Sato, Diffeomorphism groups and classification of manifolds, J. Math. Soc. Japan 21 (1969), 1-36.

[S2] ______, Diffeomorphism groups of $S^p \times S^q$ and exotic spheres, Quart. J. Math. Oxford (Ser. 2) 20 (1969), 255-276.

[S3] B. Schellenberg, The group of homotopy self equivalences of some compact CW complexes, Math. Ann. 200 (1973), 253-266.

[S4] R. Schultz, Composition constructions on diffeomorphisms of $S^p \times S^q$, Pac. J. Math. (3) 42 (1972), 739-754.

[S5] G. P. Scott, The space of homeomorphisms of a 2-manifold, Topology 9 (1970), 97-109.

[S6] W. Shih, On the group $\mathcal{E}[x]$ of homotopy equivalence maps, Bull. Am. Math. Soc. 70 (1964), 361-365.

[S7] M. Shub and D. Sullivan, Homology and dynamical systems, preprint.

[S8] L. Siebenmann, Notices Am. Math. Soc. (1967), p. 852 and p. 942.

[S9] A. J. Sieradski, Stabilization of self equivalences of the pseudo-projective spaces, Mich. Math. J. (2) 19 (1972), 109-119.

[S10] ______________, Twisted self homotopy equivalences, Pac. J. Math. 34 (1970), 789-802.

[S11] S. Smale, Diffeomorphisms of the 2 - sphere, Proc. Am. Math. Soc. 10 (1969), 621-626.

[S12] T. E. Stewart, On groups of diffeomorphisms, Proc. Am. Math. Soc. 11 (1970), 559-563.

[S13] A. G. Swarup, Pseudo-isotopies of $S^3 \times S^1$, Math. Z. 121 (1971), 201-205.

[T1] R. Tindell, Relative concordance, Topology of manifolds (Proc. Inst., Univ. of Georgia, Athens, Georgia, 1969), 453-457.

[T2] E. C. Turner, Diffeomorphisms homotopic to the identity, Trans. Am. Math. Soc. (2) 188 (1974), 1-10.

[T3] ___________, Diffeomorphisms of a product of spheres, Invent. Math. 8 (1969), 69-82.

[T4] ___________, Nilpotent diffeomorphism groups, preprint.

[T5] ___________, Rotational symmetry: basic properties and application to knot manifolds, Invent. Math. 19 (1973), 219-234.

[T6] ___________, Some finite diffeomorphism groups, Illinois J. Math, to appear.

[W1] J. B. Wagoner, Algebraic invariants for pseudo-isotopies, Proc. Liverpool Sing. Sympos. II, Springer lecture notes #209, 1971.

[W2] R. Waldhausen, On irreducible 3 - manifolds which are sufficiently large, Ann. of Math. (2) 87 (1968), 56-88.

[W3] C. T. C. Wall, Classification problems in differential topology, IV: Diffeomorphisms of handlebodies, Topology 2 (1963), 263-272.

[W4] ____________, Diffeomorphisms of 4-manifolds, J. London Math. Soc. 39 (1964), 131-140.

[W5] K. Wang, Free S^1 actions and the group of diffeomorphisms, preprint, to appear, Trans. Am. Math. Soc.

[W6] R. Wells, Concordance of diffeomorphisms and the pasting construction, Duke Math. J. (4) 39 (1972), 665-693.

SEMIFREE ACTIONS ON HOMOTOPY SPHERES

Kai Wang

This lecture will be divided into two parts. In part one we will survey the works of M. Rothenberg [5], W. Browder and T. Petrie [3] on semifree actions on homotopy spheres. In part two we will study the Atiyah-Singer invariants involved in the Rothenberg's exact sequence.

PART ONE

Our notations follow [5] closely. By G we denote always a fixed compact Lie group. A G-manifold M^n is a differentiable manifold with a fixed differentiable G action on it. By $F^k(M,G)$ we denote the k-dimensional submanifold of fixed points. An action is semifree if G acts freely outside the fixed points set. Let τ be the G-equivariant normal bundle to $F^k(M,G)$ in M^n and the action of G on each fiber is linear and thus represents an $(n-k)$-dimensional representation ρ of G. The conjugacy class of this representation is constant on the connected component of $F^k(M,G)$. Let $C(G,\rho)$ be the centralizer of $\rho(G)$ in $O(n-k)$. Then there is a reduction of the structural group of the bundle τ from $O(n-k)$ to $C(G,\rho)$. Let E be the restriction of the tangent bundle of M to $F^k(M,G)$. Then the reduction of the structural group of τ to $C(G,\rho)$ induces a reduction of the structural group of E to $C(G,\rho) \times O(k)$. A (G,ρ)-orientation on M is a further reduction of the structural group of E to the connected component of the identity of

$C(G,\rho) \times O(k)$. Note that if F is simply-connected such a reduction always exists.

Clearly, it makes sense to speak of (G,ρ) - orientation preserving equivariant diffeomorphisms. Thus we have a notion of equivalence for (G,ρ) - oriented manifolds and it is on such equivalence classes that we can define (G,ρ) - oriented connected sum in a manner formally the same as we define the ordinary connected sum [4].

Let $\rho : G \longrightarrow O(n-k)$ be a fixed point free representation. An element of $S^n(G,\rho)$ is an equivalence class of semifree (G,ρ) - oriented manifolds M^n such that M^n and $F^k(M,G)$ are homotopy spheres.

Theorem 1: (Rothenberg and Sondow [4]) For $k \geq 1$, $S^n(G,\rho)$ under the (G,ρ) - oriented connected sum is an abelian group.

Let a and a' represent elements of $S^n(G,\rho)$. We say that a is h - cobordant to a' if there is a semifree (G,ρ) - oriented manifold W which is homotopy equivalent to $S^n \times [0,1]$ and $F(W,G)$ is homotopy equivalent to $S^k \times [0,1]$ and ∂W as a (G,ρ) - oriented manifold is (G,ρ) - diffeomorphic to the dijoint union of a and -a'. It is routine to check that addition preserves h - cobordisms. Let $CS^n(G,\rho)$ be the group of h - cobordism classes of $S^n(G,\rho)$.

Theorem 2: (Rothenberg and Sondow [4]) The following sequence

$$Wh(\pi_0(G)) \xrightarrow{\phi} S^n(G,\rho) \longrightarrow CS^n(G,\rho) \longrightarrow 0$$

is exact where the map ϕ is defined as in [3]. ϕ is mono if $G = Z_m$ and n is odd.

In order to apply surgery theory to study $CS^n(G,\rho)$, another functor $RS^n(G,\rho)$ which is a variant of $CS^n(G,\rho)$ will be introduced. An element of $\overline{RS}^n(G,\rho)$ is an equivalence class of objects where an object is

(i) a (G,ρ) - oriented homotopy sphere Σ^n,

(ii) a (G,ρ) - orientation preserving embedding

$$\psi : S^k \times R^{n-k} \longrightarrow \Sigma^n$$

such that G acts freely on $\Sigma^n - (S^k \times 0)$ and G acts on $S^k \times R^{n-k}$ by

$$g(x,y) = (x, \rho(g)y).$$

An equivalence of two objects is a (G,ρ) - orientation preserving diffeomorphism

$$d : \Sigma \longrightarrow \Sigma'$$

such that the following diagram is commutative.

$$\begin{array}{ccc} & S^n \times R^{n-k} & \\ \psi \swarrow & & \searrow \psi' \\ \Sigma & \xrightarrow{\quad d \quad} & \Sigma' \end{array}$$

$\overline{RS}^n(G,\rho)$ is a group under (G,ρ) - connected sum. Define an $(n+1)$ - dimensional disk object to be a

$$\psi : D^{k+1} \times R^{n-k} \longrightarrow D^{n+1}$$

which satisfies (i) and (ii) with "sphere" replaced by "disk." Let

$RS^n(G,\rho)$ be the quotient of $\overline{RS}^n(G,\rho)$ by those elements which bound disk objects.

Let $A(G,\rho)$ be the h-space of continuous G equivariant maps of S^{n-k-1}.

<u>Theorem 3</u>: (Rothenberg [4], see also Browder and Petrie [3]) Those functors are connected by the following two exact sequences:

$$\text{(I)} \quad \cdots \longrightarrow RS^n(G,\rho) \longrightarrow CS^n(G,\rho) \xrightarrow{\alpha} \Gamma_k + \pi_k(C(G,\rho)) \xrightarrow{\mu} RS^{n-1}(G,\rho) \longrightarrow \cdots$$

$$\text{(II)} \quad \cdots \longrightarrow RS^n(G,\rho) \longrightarrow \pi_k(A(G,\rho)) \longrightarrow hS\left(S^{n-k-1}/\rho \times (D^k, S^{k-1})\right) \longrightarrow RS^{n-1}(G,\rho) \longrightarrow \cdots$$

When $G = S^1$, the normal bundle of $F(M,G)$ in M has a complex structure and the action on it is just that induced by the complex structure. Therefore $C(S^1) = U(n-k/2)$.

<u>Theorem 4</u>: (W. Browder [2]) The normal bundle of $F(\Sigma^n, S^1)$ in Σ^n is stably trivial as a complex vector bundle.

<u>Theorem 5</u>: (Browder and Petrie [3])

(i) $S^n(S^1)$ is finite if n is even,

(ii) $S^n(S^1) \otimes Q \cong \Lambda^{n,k}$ if n is odd where

$$\Lambda^{n,k} \subset H^{4*}\left(CP^{(n-k)/2-1} \times (D^{k+1}, S^k) ; Q\right)$$

is a subspace of codimension 1 if $n \equiv 1 \pmod 4$ and of codimension 0 if $n \equiv 3 \pmod 4$.

If $G = Z_2$, the action on each fiber is just the antipodal map. Hence $C(Z_2) = O(n-k)$.

Theorem 6: (see Browder and Petrie [3]).

$$S^{n-1}(Z_2) \otimes Q \cong B_{n,k} + L_n^0\left(Z_2,(-1)^{n-k}\right) \otimes Q$$

where $L_n^0\left(Z_2,(-1)^{n-k}\right)$ is the reduced Wall group and $B_{n,k} = 0$ unless $k \equiv 1 \pmod 4$ and $k \geq \frac{2}{3}(2n-1)$ in which case $B_{n,k} = Q$.

When $G = Z_m = \langle g \rangle$, $m \neq 2$, there is a set of integers $q_1, \cdots, q_\ell$ prime to m and $1 \leq q_j \leq m/2$. Then the normal bundle τ of $F(M,Z_m)$ in M splits into a Whitney sum $\bigoplus_{j=1}^{\ell} \tau_j$ of complex vector bundles so that each factor τ_j is invariant under the action of Z_m and this restriction to each fiber is just the complex multiplication by $\exp(2\pi q_j i/m)$. Let $n_j = \dim_C \tau_j$, then $\rho = \sum_{j=1}^{\ell} n_j t^{q_j}$ where t is the basic complex one dimensional representation of Z_m defined to be the multiplication by $\exp(2\pi i/m)$. In this case $C(Z_m,\rho) \cong U(n_1) \times \cdots \times U(n_\ell)$.

Theorem 7: (see Browder and Petrie [3]) For n = even,

(i) $RS^n(Z_m,\rho) \otimes Q = 0$

(ii) $RS^{n-1}(Z_m,\rho) \otimes C \cong R_{n,m} + \pi_{k-1}(O(n-k)) \otimes C$

where $R_{n,m} \subset C[Z_m]$, the complex group ring of Z_m, is the subspace generated by $\{g^j + (-1)^n g^{m-j}\}$, $j = 1,2,\cdots,[m/2]$. The map $\mu \otimes C \cong \Delta + \psi \otimes C$

where $\psi:\pi_{k-1}(C(Z_m,\rho)) \longrightarrow \pi_{k-1}(O(n-k))$ is induced by the inclusion and Δ is defined as follows: For $f:S^{k-1} \longrightarrow C(Z_m,\rho)$, let η be the vector bundle over S^k with f as characteristic map. Let Z_m act on η via ρ. Then we define

$$\Delta(f)(g^j) = \sigma(S(\eta),g^j)$$

where σ is the Atiyah-Singer invariant [1].

PART TWO

Let the complex valued functions $\Phi_r(\theta)$ be defined by the equation:

$$\Pi \frac{\tanh i\theta/2}{\tanh(x_j + i\theta)/2} = \Sigma\Phi_{i_1 \dots i_r}(\theta) C_{i_1} \cdots C_{i_r}$$

where C_i is the i-th Chern class.

For $m > 2$, let $\Lambda = \{1 \leq k < m/2 \mid (k,m) = 1\}$ and let $\lambda = |\Lambda|$. Consider the following $\lambda \times \lambda$ matrix

$$\tilde{\Phi}_r^{(m)} = \left(\Phi_r\left(\frac{4k\pi}{m}\right)\right)_{j,k \in \Lambda}.$$

This matrix plays the most important role in the formula of G-signatures [1] especially when the fixed points set is a sphere.

Lemma 1:

$$\Phi_{2m}(\theta) = -(-i)^{2m} \frac{\cos \theta}{\sin^{2m}\theta} P(\cos^2\theta)$$

$$\Phi_{2m+1}(\theta) = -(-i)^{2m+1} \frac{1}{\sin^{2m}(\theta)} Q(\cos^2\theta)$$

where P and Q are some polynomials with non-negative coefficients.

For simplicity in the following unless otherwise stated we will always assume that m is an odd prime and $m \not\equiv 1 \pmod 8$.

For $k \in N$, let $\alpha(k)$, $\delta(k)$ be integers such that

(i) $1 \leq \alpha(k) \leq m/2$,

(ii) either $\alpha(k) \equiv k \pmod m$ or $\alpha(k) + k \equiv 0 \pmod m$,

(iii) $0 \leq \delta(k) \leq 1$,

(iv) let $k = sm + t$ with $0 \leq t < m$, then

$\delta(k) = 0$ if $0 \leq t \leq m/2$, $\delta(k) = 1$ otherwise.

<u>Lemma 2</u>: There is an $x \in \Lambda$ such that $\alpha(x^{\lambda}) = 1$ and $\alpha(x^j) \neq 1$ for $0 < j < \lambda$.

For $1 \leq s \leq \lambda$, let $\rho(s) = \alpha(x^s)$. It is easy to check that ρ is a one-to-one correspondence from $\{1,2,\cdots,\lambda\}$ to Λ. For $k \in \Lambda$, let $\phi(k) = \lambda - \rho^{-1}(k)$.

Let $\eta(x)$ be a solution of the following equation

$$\lambda y \equiv 2 \sum_{j \in \Lambda} \delta(jx) \pmod 4$$

Then, for $1 \leq m < \lambda$, define

$$\eta(\alpha(x^m)) \equiv \eta(x) + \eta(\alpha(x^{m-1})) + 2\delta(x \cdot \alpha(x^{m-1})) \pmod 4$$

by induction on m.

Lemma 3: For $j,k \in \Lambda$,

(i) $\phi(j) + \phi(k) \equiv \phi(\alpha(jk)) \pmod{\lambda}$

(ii) $\eta(j) + \eta(k) \equiv \eta(\alpha(jk)) + 2\delta(jk) \pmod{4}$

Theorem 1. There are integers $0 \leq \phi(j) \leq \lambda - 1$, $0 \leq \eta(j) \leq 3$ such that

$$\text{(i)}\quad \det \tilde{\Phi}^{(m)}_{2r} = \pm \prod_{s=0}^{\lambda-1} \Big(\sum_{j \in \Lambda} \Phi_{2r}\big(\tfrac{2j\pi}{m}\big)\omega^{s\phi(j)} \Big),$$

$$\text{(ii)}\quad \det \tilde{\Phi}^{(m)}_{2r+1} = \pm \prod_{s=0}^{\lambda-1} \Big(\sum_{j \in \Lambda} \Phi_{2r+1}\big(\tfrac{2j\pi}{m}\big)\omega^{s\phi(j)} \big(\sqrt{-1}\big)^{\eta(j)} \Big)$$

where ω is the λ-th primitive root of unity.

Condition A: $m \leq 7$ or $r \geq \dfrac{\log (\lambda - 1) \sec \frac{\pi}{m}}{\log 2}$

Lemma 4: Condition A implies $\det \tilde{\Phi}^{(m)}_{r} \neq 0$.

Conjecture: $\det \tilde{\Phi}^{(m)}_{r} \neq 0$.

Let (Σ^{2n}, Z_m) be a semifree action on a homotopy sphere Σ^{2n} with fixed point set F^{2r}, a homotopy sphere. The normal bundle τ of F^{2r} in Σ^{2n} splits equivariantly as Whitney sum $\tau = \bigoplus_{j \in \Lambda} \tau_j$ of complex vector bundles as in Part I. Since Σ^{2n} has no middle dimensional cohomology classes, $\operatorname{Sign}(\Sigma^{2n}, g^k) = 0 \;\forall\, k$. By Atiyah-Singer G-signature Theorem [1],

$$\text{Sign}\,(\Sigma^{2n},g^k) = K_k\Big(\sum_{j\,\varepsilon\,\Lambda} \Phi_r\big(\tfrac{2jk\pi}{m}\big)c_r(\tau_j)\Big) \cap [F^{2r}] \quad \text{for } j \,\varepsilon\, \Lambda$$

where K_k are constants. If Φ_r is invertible, this implies that $c_r(\tau_j) = 0$. Hence we proved the following:

Theorem 2: Let (Σ^{2n},Z_m) be a semifree action as above. If m and r satisfy the Condition A, the normal bundle of F^{2r} in Σ^{2n} is equivariantly stably trivial.

This result has also been proved by R. Schultz.

Corollary 1: Let m and r be as in Condition A. If n is even, $CS^n(Z_m,\rho)$ is finite and

$\text{rank}\; CS^{n-1}(Z_m,\rho) \otimes C =$

$\text{rank}\; R_{n,m} + \text{rank}\; \pi_{r-1}(O(n-r)) \otimes C - \text{rank}\; \pi_r(C(Z_m,\rho)) \otimes C.$

When $m = 4k$, we consider $\rho = n_1 t + n_2 t^{2k-1}$ where $n_1 \geq r$, $n_2 \geq r$. $C(Z_{4k},\rho) \cong U(n_1) \times U(n_2)$. It is easy to show that $\dim_C \ker \Delta = 1$.

Corollary 3: There exist infinitely many inequivalent semifree Z_{4k} actions on S^{2n} with fixed point set S^{2r} for $n \geq 3r$ and r odd.

This is the first known infinite family of semifree actions on even dimensional sphere. Similarly we can prove:

Corollary 4: For $n \geq 5r$, $k \geq 4$, there are infinitely many inequivalent semifree Z_{4k} actions on S^{2n} with fixed point set S^{2r}.

Remark 1: If r is odd, Theorem 2 and Corollary 1 are true without assuming $m \not\equiv 1 \pmod 8$.

Remark 2: It is clear that the result announced in [7] is in general false.

REFERENCES

1. M. F. Atiyah and I. M. Singer, The index of elliptic operators III, Ann. of Math. (2) 87 (1968), 346-604.

2. W. Browder, Surgery and the theory of differentiable transformation groups, Proceedings of Conference on Transformation Groups (New Orleans, 1967), Springer-Verlag 1968, 1-46.

3. W. Browder and T. Petrie, Diffeomorphisms of manifolds and semifree actions on homotopy spheres, Bull. A.M.S. 77 (1971), 160-163.

4. M. Rothenberg and J. Sondow, Non-linear smooth representations of compact Lie groups (preprint).

5. M. Rothenberg, Differentiable group actions on spheres (preprint).

6. R. Schultz, Rational h - cobordism invariants for lens space bundles (preprint).

7. K. Wang, Differentiable actions on 2n - spheres, Bull. A.M.S. 78 (1972), 971-973.

Cell-Like Mappings

by

T. A. Chapman[1]

§1. Introduction. The purpose of this paper is to establish the following result which implies the topological invariance of Whitehead torsion for finite CW complexes.

CE Mapping Theorem. *If* X *and* Y *are finite* CW *complexes and* $f : X \longrightarrow Y$ *is a cell-like map, then* f *is simple (i.e.* f *is a simple homotopy equivalence).*

Recall from [5] that a map is *cell-like* (or CE) provided that it is proper, onto, and each point-inverse is cell-like (i.e. it can be embedded in some euclidean space as a cellular set).

The first proof of the CE Mapping Theorem was given in [1] and used Hilbert cube manifold theory. More recently another proof has been given by R. D. Edwards which uses n-manifold theory and relies upon his notion to TOP regular neighborhoods [2]. Both proofs are inspired by the techniques of Siebenmann's Approximation Theorem [9]. The proof we give of the CE Mapping Theorem, while inspired by the techniques of [9], avoids both Hilbert cube manifold theory and the tools of Edwards. It relies instead upon results from the noncompact simple homotopy theory of Siebenmann [7].

We leave some rather difficult-looking questions open.

CE Image Question. *If* X *is a compact finite-dimensional* AR, *then is* X *the* CE *image of some n-cell?*

---

[1]The author is an A. P. Sloan Fellow and is partially supported by NSF Grant GP-28374.

This should be of some interest because of the following result (see §6): An affirmative answer to the CE Image Question would imply that every compact finite-dimensional ANR is the CE image of some compact polyhedron. Therefore an affirmative answer to the CE Image Question would imply that every compact finite-dimensional ANR has finite homotopy type, a desirable result [8].

Along the same lines we pose the following question [1].

CE Classification Question. If X ($\dim X < \infty$) is the CE image of compact polyhedra Y and Z, then must Y and Z be simple homotopy equivalent?

The results of Lacher [5] imply that Y and Z would have to be homotopy equivalent, but the question of simple homotopy equivalence seems much more delicate. Such a question would probably have to be dealt with in any program of extending simple homotopy theory to the class of all compact ANR's (see [1]).

We remark that throughout this paper all spaces will be locally-compact, finite-dimensional, separable, and metrizable. We will freely use results from Lacher [5] concerning CE maps and from Siebenmann [7] concerning simple homotopy equivalences.

The author is grateful to L. C. Siebenmann for some valuable comments on an earlier version of this paper.

§2. The Main Lemma. In this section we establish the main step in the proof of the CE Mapping Theorem. For notation let R^n denote euclidean n-space and let $B_r^n = [-r, r]^n \subset R^n$ denote the n-ball of radius r.

Main Lemma. If X is a polyhedron and $f : X \longrightarrow R^n$ is a CE map, then there exists a polyhedron Y and a CE map $g : Y \longrightarrow R^n$ such that g is a PL homeomorphism over a neighborhood of ∞ and $g = f$ over a neighborhood of 0.

Remark. When we say that g is a PL homeomorphism over a neighborhood of ∞ we mean that there exists some r sufficiently large so that the restriction $g| : g^{-1}(R^n - B^n_r) \longrightarrow R^n - B^n_r$ is a PL homeomorphism. Similarly when we say that $g = f$ over a neighborhood of 0 we mean that there exists some r sufficiently small so that $g^{-1}(\mathrm{Int}(B^n_r)) = f^{-1}(\mathrm{Int}(B^n_r))$ and $g|g^{-1}(\mathrm{Int}(B^n_r)) = f|f^{-1}(\mathrm{Int}(B^n_r))$.

Proof. We use $e : R \longrightarrow S^1$ for the covering projection defined by $e(x) = \exp(\pi i x/4)$. Then $e^n : R^n \longrightarrow T^n$ is the product covering projection defined by $e^n = e \times e \times \cdots \times e$, where $T^n = S^1 \times S^1 \times \cdots \times S^1$ is the n-torus. Let T^n_0 be the punctured torus and let $\alpha : T^n_0 \longrightarrow R^n$ be a PL immersion such that $\alpha e^n | B^n_3 : B^n_3 \longrightarrow B^n_3$ is the identity (see [3], p. 48 for references). We are going to work our way through the accompanying diagram of spaces and maps. The knowledgeable reader will note the similarity between this diagram and the diagram of [9].

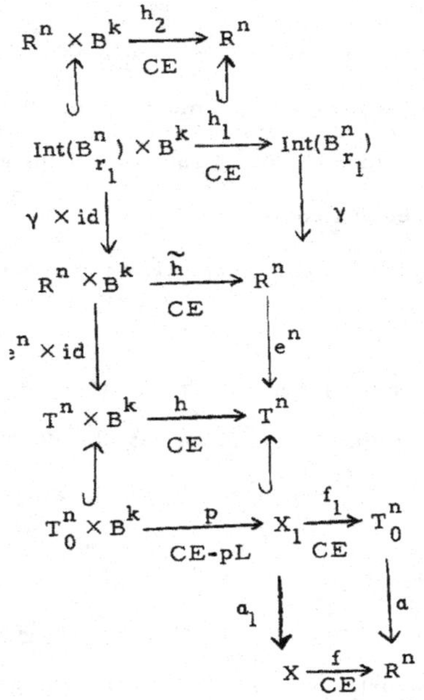

1. Construction of X_1. We define X_1 to be the pull-back:

$$X_1 = \{(x, t) \in X \times T_0^n | f(x) = \alpha(t)\} .$$

Then $\alpha_1 : X_1 \longrightarrow X$, $f_1 : X_1 \longrightarrow T_0^n$ are projection maps. It is easy to check that α_1 is an immersion and f_1 is CE. Thus X_1 inherits a PL structure from X making X_1 a polyhedron and α_1 a PL map. We also note that $\alpha_1 | f_1^{-1} e^n(B_3^n)$ is 1-1.

2. Construction of p. It follows from [5] that f_1 is a proper homotopy equivalence and it follows from [7] that f_1 is simple. As observed in [7], this implies that X_1 and T_0^n have closed regular neighborhoods in an euclidean space of large dimension which are PL homeomorphic. As T_0^n has regular neighborhoods PL homeomorphic to $T_0^n \times B^k$, this implies that there is a PL homeomorphism of $T_0^n \times B^k$ onto M, a regular neighborhood of X_1. Then p is obtained by composing this PL homeomorphism with a PL collapse of M onto X_1. Moreover we can require that $f_1 p : T_0^n \times B^k \longrightarrow T_0^n$ is proper homotopic to the projection map of $T_0^n \times B^k$ onto T_0^n.

3. Construction of h. The map h just extends $f_1 p$ by sending $t_0 \times B^k$ to t_0, where $T^n = T_0^n \cup t_0$. Note that h is CE and h is homotopic to the projection map.

4. Construction of $\widetilde{h}$. The map $\widetilde{h}$ covers h. It follows from elementary covering space theory that $\widetilde{h}$ is CE and $\widetilde{h}$ is a bounded distance from the projection map of $R^n \times B^k$ onto R^n. It is also easy to see that $(e^n \times id) | \widetilde{h}^{-1}(B_3^n)$ is 1-1 and onto $h^{-1}(B_3^n)$.

5. Construction of γ. Choose $r_1 > r \geq 3$ large enough so that $B^n_r \times B^k$ contains $\tilde{h}^{-1}(B^n_3)$. Let $\gamma : \mathrm{Int}(B^n_{r_1}) \longrightarrow R^n$ be a radially defined PL homeomorphism such that $\gamma = \mathrm{id}$ on B^n_r. Then define $h_1 : \mathrm{Int}(B^n_{r_1}) \times B^k \longrightarrow \mathrm{Int}(B^n_{r_1})$ to make the appropriate rectangle commute. Note that h_1 is CE.

6. Construction of h_2. The map h_2 extends h_1 by defining h_2 to be the projection on $(R^n - \mathrm{Int}(B^n_{r_1})) \times B^k$. Note that h_2 is a CE map and $h_2 = f\alpha_1 p(e^n \times \mathrm{id})$ over $\mathrm{Int}(B^n_3)$. This latter fact is easily seen because $\alpha e^n | B^n_3 = \mathrm{id}$.

Our required $g : Y \longrightarrow R^n$ is now constructed from $h_2 : R^n \times B^k \longrightarrow R^n$ as follows. We first modify $h_2 : R^n \times B^k \longrightarrow R^n$ by using the projection map of $R^n \times B^k$ onto R^n to identify $(R^n - \mathrm{Int}(B^n_{r_1})) \times B^k$ with $R^n - \mathrm{Int}(B^n_{r_1})$. This gives a CE map g_1 of a polyhedron Y_1 onto R^n such that g_1 is a PL homeomorphism over a neighborhood of ∞ and $g_1 = f\alpha_1 p(e^n \times \mathrm{id})$ over $\mathrm{Int}(B^n_3)$. But $\alpha_1 p(e^n \times \mathrm{id}) | : g_1^{-1}(\mathrm{Int}(B^n_3)) \longrightarrow f^{-1}(\mathrm{Int}(B^n_3))$ is a PL collapse. We use this collapse to identify a neighborhood of $g_1^{-1}(0)$ in Y_1 with a neighborhood of $f^{-1}(0)$ in X. This then yields a polyhedron Y and a CE map $g : Y \longrightarrow R^n$ which fulfills our requirements.

§3. The Main Theorem. In this section we use the inversion trick of [9] to interchange the roles of 0 and ∞ in the statement of the Main Lemma.

Main Theorem. *If X is a polyhedron and $f : X \longrightarrow R^n$ is a CE map, then there exists a polyhedron Y and a CE map $g : Y \longrightarrow R^n$ such that g is a PL homeomorphism over a neighborhood of 0 and $g = f$ over a neighborhood of ∞.*

Proof. Using the Main Lemma let Y_1 be a polyhedron and let $g_1 : Y_1 \longrightarrow R^n$ be a CE map such that g_1 is a PL homeomorphism over a neighborhood of ∞ and $g_1 = f$ over $\mathrm{Int}(B_3^n)$. If $\tilde{Y}_1$ is the one-point compactification of Y_1, then $\tilde{Y}_1$ is a polyhedron and g_1 extends to a CE map

$$\tilde{g}_1 : \tilde{Y}_1 \longrightarrow S^n = R^n \cup \infty.$$

Let $Y_2 = \tilde{Y}_1 - \tilde{g}_1^{-1}(0)$ and let $g_2 : Y_2 \longrightarrow S^n - 0$ be given by restricting $\tilde{g}_1$. Applying the Main Lemma once again we get a polyhedron Y_3 and a CE map $g_3 : Y_3 \longrightarrow S^n - 0$ such that g_3 is a PL homeomorphism over $\mathrm{Int}(B_1^n) - 0$ and $g_3 = g_2$ over $S^n - B_2^n$. Then

$$g_3| : g_3^{-1}(\mathrm{Int}(B_3^n)-0) \longrightarrow \mathrm{Int}(B_3^n) - 0$$

gives a CE map which is a PL homeomorphism over $\mathrm{Int}(B_1^n) - 0$ and which equals f over $\mathrm{Int}(B_3^n) - B_2^n$.

Our required $g : Y \longrightarrow R^n$ is now obtained by modifying $g_3|$. We first add to $g_3^{-1}(\mathrm{Int}(B_3^n)-0)$ the one-point compactification of $g_3^{-1}(B_1^r-0)$. This gives a polyhedron Y_4 and a CE map $g_4 : Y_4 \longrightarrow \mathrm{Int}(B_3^n)$ such that g_4 is a PL homeomorphism over a neighborhood of 0 and $g_4 = f$ over $\mathrm{Int}(B_3^n) - B_2^n$. Then Y is obtained from Y_4 by adding $X - f^{-1}(B_2^n)$ to Y_4 and g is obtained from g_4 by a trivial extension.

We will need the following corollary of the Main Theorem in the proof of the CE Mapping Theorem. The proof is easy and is accordingly omitted.

Corollary. Assuming the notation of the Main Theorem we can find compact subpolyhedra $A \subset Y$, $B \subset X$, and a map $h : Y \longrightarrow X$ such that $h|Y - \mathrm{Int}(A) : Y - \mathrm{Int}(A) \longrightarrow X - \mathrm{Int}(B)$ is the identity and $h|A : A \longrightarrow B$ is a homotopy equivalence.

§4. The Polyhedral CE Mapping Theorem. In this section we show how to use the Corollary of the Main Theorem to prove the CE Mapping Theorem for polyhedra.

Polyhedral CE Mapping Theorem. If X and Y are compact polyhedra and $f : X \longrightarrow Y$ is a CE map, then f is simple.

Proof. We induct on dim Y. The theorem is clearly true for dim Y = 0, as X would then be the union of a finite collection of pairwise disjoint, compact, contractible subpolyhedra. Passing to the inductive step write $Y = Y^{n-1} \cup (\bigcup_{i=1}^{k} \alpha_i(R^n))$, where $\alpha_i : R^n \longrightarrow Y$ is a PL open embedding, $\{\alpha_i(R^n)\}_{i=1}^{k}$ is the collection of top dimensional open n-cells, and Y^{n-1} is the (n-1)-skeleton of Y. Applying the Main Theorem k times there exists a polyhedron Z and a PL map $g : Z \longrightarrow Y$ such that g is a PL homeomorphism over $\bigcup_{i=1}^{k} \alpha_i(B_1^n)$ and $g = f$ over $Y - \bigcup_{i=1}^{k} \alpha_i(B_2^n)$.

We now show that g is simple. Our proof uses the Sum Theorem [7]. Let $Y_1 = \bigcup_{i=1}^{k} \alpha_i(B_1^n)$, $Y_2 = Y - \text{Int}(Y_1)$, $Y_0 = Y_1 \cap Y_2$, and let $Z_j = g^{-1}(Y_j)$ for $j = 0, 1, 2$. The Z_j's are subpolyhedra of Z and the Y_j's are subpolyhedra of Y. If $g_j : Z_j \longrightarrow Y_j$ is defined by restricting g, then g_0 and g_1 are simple (as they are PL homeomorphisms). To see that g_2 is simple let $r : Y_2 \longrightarrow Y^{n-1}$ be a CE collapse (which is simple) and apply the inductive hypothesis to conclude that $rg_2 : Z_2 \longrightarrow Y^{n-1}$ is simple. It then follows that g_2 is simple. Then g is simple by the Sum Theorem.

We now apply the Corollary of the Main Theorem k times. This means that we can choose compact subpolyhedra $A_i \subset g^{-1}\alpha_i(R^n)$, $B_i \subset f^{-1}\alpha_i(R^n)$ such that if $A = \bigcup_{i=1}^{k} A_i$ and $B = \bigcup_{i=1}^{k} B_i$, then there exists a map $h : Z \longrightarrow X$ such that $h|Z - \text{Int}(A) : Z - \text{Int}(A) \longrightarrow X - \text{Int}(B)$ is the identity and $h|A : A \longrightarrow B$ is a homotopy equivalence. It is easy to check that fh is homotopic to g. Therefore f is simple iff h is simple. But it follows from the Sum Theorem that h is simple (because $h|A : A \longrightarrow B$ has trivial torsion in Wh(X)).

A corollary of the preceeding proof gives us a more general result.

Corollary. *If X is a compact polyhedron, Y is a finite CW complex, and $f : X \longrightarrow Y$ is a CE map, then f is simple.*

§5. *The CE Mapping Theorem.* In this section we prove the CE Mapping Theorem as stated in §1. We will first need the following result of R. D. Edwards. Our proof is somewhat different.

Lemma 5.1. *Every finite CW complex is the CE image of some polyhedron.*

Proof. Inducting on the number of cells in the CW complex it clearly suffices to establish the following result: *If X is a compactum which is the CE image of some polyhedron and X_1 is obtained from X by attaching an n-cell, then X_1 is also the CE image of some polyhedron.*

We first show that X is a CE retract of some polyhedron. Let P be a polyhedron and let $r : P \longrightarrow X$ be a CE map. It follows from [6] that r can be realized in some euclidean space. This means that we can regard P and X as subsets of some euclidean space R^n, with P polyhedral in R^n, and we can extend r to a map $\tilde{r} : R^n \longrightarrow R^n$ such that $\tilde{r}$ gives a homeomorphism of $R^n - P$ onto $R^n - X$. Let M be a regular neighborhood of P in R^n and let $\alpha : M \longrightarrow P$ be a CE retraction. Then $\tilde{r}(M)$ is homeomorphic to M (by [9]), therefore $\tilde{r}(M)$ is a polyhedron. One could also deduce this directly by using Sher's ideas [6] without appealing to [9]. We can define a CE retraction $\tilde{\alpha} : \tilde{r}(M) \longrightarrow X$ by $\tilde{\alpha} = r\alpha\tilde{r}^{-1}$.

Thus we may assume that we have a CE retraction $r : M \longrightarrow X$, for some compact PL manifold M containing X in its interior. We also have $X_1 = X \cup_f B_1^n$, where $f : \mathrm{Bd}(B_1^n) \longrightarrow X$ is the attaching map. For $k > n$ let $g : \mathrm{Bd}(B_1^n) \longrightarrow M \times B_2^k$ be defined by $g(x) = (f(x), x)$ (where $B_1^n \equiv B_1^n \times 0 \subset B_2^k$) and put $Y = (M \times B_2^k) \cup_g B_1^n$. For k large enough,

Y is a polyhedron. By collapsing the B_2^k -factor in Y to 0 we get a CE map of Y onto $M \cup_f B_1^n$. Then using the CE retraction r we get a CE map of $M \cup_f B_1^n$ onto X_1.

Proof of the CE Mapping Theorem. We are given a CE map $f : X \longrightarrow Y$, where X and Y are finite CW complexes. Using Lemma 5.1 there exists a polyhedron P and a CE map $g : P \longrightarrow X$. Using the Corollary of the Polyhedral CE Mapping Theorem it follows that g and fg are simple, therefore f is simple.

§6. CE images of polyhedra. In this section we concern ourselves with the question of determining the set of all compacta which are CE images of polyhedra (denoted $\mathcal{CE}$). We have just shown (Lemma 5.1) that any finite CW complex is in $\mathcal{CE}$ and it follows from the work of Kirby-Siebenmann [4] (concerning finiteness of homotopy types) that any compact n-manifold is in $\mathcal{CE}$. More generally one could probably establish a "handle version" of the Main Theorem and use this to prove that any locally triangulable compact metric space is in $\mathcal{CE}$. Work of Lacher [5] implies that any space in $\mathcal{CE}$ must be a compact ANR, but there is no known counterexample to the converse of this statement. We can reduce the problem somewhat.

Theorem 6.1. *If every compact AR lies in $\mathcal{CE}$, then every compact ANR also lies in $\mathcal{CE}$.*

Proof. If X is a compact ANR, then we can attach a finite number of cells to obtain an AR. Therefore, the proof of our proposition reduces to establishing the following result: *If X_1 is obtained from X by attaching a cell and $X_1 \in \mathcal{CE}$, then $X \in \mathcal{CE}$.*

Write $X_1 = X \cup_f B_1^n$, where $f : Bd(B_1^n) \longrightarrow X$ is the attaching map. We are given a polyhedron P and a CE map $r : P \longrightarrow X_1$. Applying the Main Theorem there exists a polyhedron Q and a CE map $s : Q \longrightarrow X_1$ such that s is a PL homeomorphism over $B_{1/2}^n$ (where we regard X_1 as the disjoint union of X and $Int(B_1^n)$ along with an appropriate topology). Put $X_2 = X_1 - Int(B_{1/2}^n)$ and note that $X_2 \in \mathcal{CE}$. But X is the CE image of X_2, therefore $X \in \mathcal{CE}$.

References

1. T. A. Chapman, Cell-like mappings of Hilbert cube manifolds: Applications to simple homotopy theory, Bull. Amer. Math. Soc., to appear.
2. R. D. Edwards, TOP regular neighborhoods, handwritten manuscript.
3. R. C. Kirby, Lectures on triangulation of manifolds, UCLA, 1969.
4. R. C. Kirby and L. C. Siebenmann, On the triangulation of manifolds and the Hauptvermutung, Bull. Amer. Math. Soc. 75(1969), 742-749.
5. C. Lacher, Cell-like mappings I, Pacific J. of Math. 30(1969), 717-731.
6. R. B. Sher, Realizing cell-like maps in euclidean space, General Top. and its App. 2(1972), 75-89.
7. L. C. Siebenmann, Infinite simple homotopy types, Indag. Math. 32(1970), 479-495.
8. ________________, On the homotopy type of compact topological manifolds, Bull. Amer. Math. Soc. 74(1968), 738-742.
9. ________________, Approximating cellular maps by homeomorphisms, Topology, to appear.

ON INTEGRAL CURRENTS AND THE DOLD-THOM CONSTRUCTION

by

Ross Geoghegan[†]

§1. Introduction

The concept of 0-dimensional singular chain, or formal finite sum of points in a topological space, is one of the simplest in algebraic topology. Yet this concept links algebraic topology with integration theory, the calculus of variations and Plateau's Problem, in a manner which does not appear to be widely known among topologists. The link occurs when one endows $\Delta_0(X)$, the group of 0-dimensional singular chains in a space X, with a topology, making it a topological abelian group.

Two topologies occur in the literature. One of them, studied by Dold and Thom [2] makes $\Delta_0(X)$ into a CW complex when X is reasonable. With this topology, $\Delta_0(X)$ is usually known as $AG(X)$, and can be thought of as the free topological abelian group generated by the space X. This is the "Dold-Thom Construction" to which the title

[†]Supported in part by National Science Foundation Grant P038761.

refers. Their theorem, Theorem 2.1 below, states that the homotopy groups of $AG(X)$ are naturally isomorphic to the homology groups of X. This gives a beautifully simple way of constructing Eilenberg-MacLane spaces and certain classifying spaces (2.4 below), and it helps one understand the place of quasifibrations in algebraic topology. Their method of proof is algebraic: essentially they show that the composition $\pi_* \circ AG$ of the functor AG with the homotopy group functor is a homology theory, and that homology theories are unique up to isomorphism. (An alternative semi-simplicial proof, due to Puppe [11] and described in §3 below, shows that the Dold-Thom Theorem is connected to a theorem of Moore which expresses the homotopy groups of a simplicial group complex as the homology groups of a certain chain complex.)

The other topology on $\Delta_0(X)$ makes it a metric space with the homotopy type of a CW complex. With this topology, $\Delta_0(X)$ is usually known as $I_0(X)$, the space of 0-dimensional integral currents in X. It is studied by Federer and Fleming [4] and Almgren [1] and has roots in Whitney's book [13]. The identity map "1": $AG(X) \longrightarrow I_0(X)$ is continuous and is a homotopy equivalence (for suitable X). Thus the homotopy groups of $I_0(X)$ are the homology groups of X. But to put it that way is to obscure the point. The <u>proof</u> of this "Dold-Thom Theorem for $I_0(X)$" given by Almgren in [1] (or, equivalently, the proof that "l" is a weak homotopy equivalence) uncovers the geometry underlying the algebra of Dold and Thom. It links algebraic topology with integration theory by explaining, on the level of cycles, why the Dold-Thom Theorem is true, and it leads Almgren to generalizations (stated in Theorem 6.1 below) to which the algebra alone could never have led.

But Almgren's proof is partly analytical. He draws on the theory of integral currents, developed by Federer and Fleming [4], and thus on a substantial amount of measure theory. This, together with the considerable technical difficulty of his proof, led the author to look for a strictly topological proof which would preserve Almgren's geometrical insights. Such a proof is sketched in §4 below.

The purpose of this paper is to describe all these ideas. Apart from §4, the work discussed here is all in the literature (though not all in papers on "topology"). Proofs of the main theorems are sketched, and enough definitions are given to make the paper readable on its own. No background in integration theory is needed. The most topological, least analytical, way of looking at things has always been used. The plan is to start with algebraic topology, to move on to geometric topology, and to end with geometric integration theory. The Dold-Thom Theorem is discussed in §2, and

Puppe's semi-simplicial proof is sketched in §3. The author's geometrical proof is outlined in §4. Whitney's flat norm is described in §5 in preparation for §6, where integral currents are defined and Almgren's Theorem is stated. As a digression, Fleming's integral current solution of Plateau's Problem is stated in §7, because it is so interesting, and can be stated easily at that stage. §8 contains some observations on the roles of topology and analysis in the context of the paper, as well as some conjectures concerning the integral current groups as infinite-dimensional manifolds.

Terminology. For reference we list here some of our terminology and definitions. The integers will be denoted by $\mathbb{Z}$, the real numbers by $\mathbb{R}$, euclidean n-space by $\mathbb{R}^n$. A polyhedron will be the geometric realization of an abstract simplicial complex: it will carry the weak topology. A subpolyhedron will be the geometric realization of a subcomplex. An absolute neighborhood retract (abbreviation ANR) will be a metrizable space which is a neighborhood retract of any metrizable space in which it is homeomorphically embedded. A compact Lipschitz neighborhood retract (abbreviation CLNR) will be a compact metric space (with chosen metric) which is a Lipschitz neighborhood retract of any metric space in which it is bi-Lipschitz homeomorphically embedded. CLNR's need not be finite-dimensional. In spite of appearances, this definition agrees with the usual one ([1], [3], [4]) for finite-dimensional CLNR's.

§2. The Theorem of Dold and Thom

The main theorem is 2.1. The main idea is in Proposition 2.3, from which 2.1 follows in the manner indicated. Some consequences are summarized in Remark 2.4. Interesting generalizations of the construction can be found in McCord's paper [10].

Let X be a Hausdorff space with base point $*$. For each positive integer q, the group of permutations of $(1,\cdots,q)$ acts on the q-fold product X^q by permuting coordinates. Let the quotient space of this action be denoted by $SP^q(X)$, the q-fold symmetric product of X. The image in $SP^q(X)$ of $(x_1,\cdots,x_q)$ in X^q is denoted by $[x_1,\cdots,x_q]$. There is an "inclusion" $SP^q(X) \subset SP^{q+1}(X)$ defined by $[x_1,\cdots,x_q] \longmapsto [*,x_1,\cdots,x_q]$. Let $SP(X,*) = \varinjlim SP^q(X)$, the direct limit being taken with respect to inclusions. $SP(X,*)$ is called the infinite symmetric product of $(X,*)$.

$SP(X,*)$ admits an addition

$\alpha([x_1,\cdots,x_q],[y_1,\cdots,y_r]) = [x_1,\cdots,x_q,y_1,\cdots y_r]$ under which it becomes a free abelian monoid with $*$ as the zero element. In fact ([2] Theorem 3.9) if X is a separable polyhedron then $SP(X,*)$ is also a topological monoid.

SP is a reasonable functor. If $f:(X,*) \longrightarrow (X',*')$ is a continuous function between pointed Hausdorff spaces then the obvious induced homomorphism $SP(f):SP(X,*) \longrightarrow SP(X',*')$ is continuous. If g is pointedly homotopic to f, $SP(f)$ is homotopic to $SP(g)$ in the obvious manner.

Consider now the space $X \vee X = \{(x_1,x_2) \in X \times X | x_1$ or $x_2 = *\}$ (abbreviating its base point $(*,*)$ to $*$). Consider also the free abelian group $AG(X,*)$ generated by the set $X\setminus\{*\}$. It is convenient to embed X in $AG(X,*)$, identifying each $x \in X\setminus\{*\}$ with $+x \in AG(X,*)$ and $*$ with the zero element. Consider the function $\tau:X \vee X \longrightarrow AG(X,*)$ defined by $\tau(x,*) = +x$ and $\tau(*,x) = -x$. Since $SP(X \vee X,*)$ is a free abelian monoid, τ extends naturally to a monoid-homomorphism $\eta:SP(X \vee X,*) \longrightarrow AG(X,*)$. η is surjective.

Endow $AG(X,*)$ with the quotient topology induced by η. In fact

$AG(X,*) = \varinjlim_q \eta(SP^q(X \vee X))$ (see 4.3 of [2]). If X is a separable

polyhedron $AG(X,*)$ becomes a topological group (see 4.8 of [2]).

Our remarks about the functorial properties of SP apply equally to AG (see 4.5 of [2]).

It is convenient to define AG on unpointed spaces too. If X is a space, adjoin a discrete base point $*$ and abbreviate $AG(X \cup \{*\})$ to $AG(X)$. In the language of category theory, $AG(X)$ is the free Hausdorff topological abelian group generated by the Hausdorff space X.

Here is the theorem of Dold and Thom.

<u>Theorem 2.1</u>: (see 6.10, I, of [2]). <u>If X is a separable polyhedron, then the singular homology group $H_q(X;\mathbb{Z})$ with integer coefficients ($q \geq 0$) is isomorphic to $\pi_q(AG(X))$. In fact, on the category of separable polyhedra and piecewise linear maps, the functors $H_*(\cdot;\mathbb{Z})$ and $\pi_* \circ AG$ are naturally equivalent.</u>

<u>Remark 2.2</u>: An equivalent version says that if $* \in X$, $\pi_q(AG(X,*))$ is isomorphic to the reduced singular homology group $\tilde{H}_q(X;\mathbb{Z})$.

Theorem 2.1 depends principally on the following.

<u>Proposition 2.3</u>: ([2], 5.4) *Let* $(X,*)$ *be a separable pointed polyhedron, let* A *be a subpolyhedron of* X *which contains* $*$, *let* $p:X \longrightarrow X/A$ *be the quotient map and let the quotient point of* X/A *be* $*'$. *Then* $AG(p):AG(X,*) \longrightarrow AG(X/A,*')$ *is the projection of a principal fiber bundle with fiber and structure group* $AG(A,*)$.

Proposition 2.3 is proved by constructing a section in a neighborhood of $*'$: one works by induction on q, defining the section on a neighborhood of $*'$ in each $\eta(SP^q(X/A \vee X/A))$.

Actually, in what follows one does not need the full strength of Proposition 2.3. One need only know that the map $AG(p)$ is a quasi-fibration; i.e., that

$$AG(p)_{\#}:\pi_*(AG(X,*),AG(A,*)) \longrightarrow \pi_*(AG(X/A,*'))$$

is a natural isomorphism.

Theorem 2.1 is then proved as follows. Take X and adjoin a discrete base point $*$ to make $X' = X \cup \{*\}$. $AG(X) \equiv AG(X',*)$. $H_q(X;\mathbb{Z})$ is isomorphic to the relative group, $H_q(X',\{*\};\mathbb{Z})$. One then shows that $\pi_q(AG(X',*))$ and $H_q(X',\{*\};\mathbb{Z})$ are isomorphic by showing that on the category of pointed separable polyhedra both functors satisfy Puppe's Axioms for a Homology Theory. A uniqueness theorem for this axiom system then gives the required result.

The only non-trivial axioms to be checked are:

(a) there is a natural equivalence $(q \geq 0)$ between $\pi_q \circ AG$ and $\pi_{q+1} \circ AG \circ \Sigma$, Σ denoting reduced suspension;

(b) if C_f is the reduced mapping cone of $f:(X,*) \longrightarrow (Y,*')$, then

$$\pi_q(AG(X,*)) \xrightarrow{AG(f)_\#} \pi_q(AG(Y,*')) \xrightarrow{AG(i)_\#} \pi_q(AG(C_f,*'))$$

is exact $(q \geq 0)$, $i:(Y,*') \longrightarrow (C_f,*')$ denoting inclusion.

In fact (a) and (b) are verified by using the quasi-fibration property given by 2.3 for the special cases

$$(X,*) \longrightarrow (CX,*) \longrightarrow (\Sigma X,*)$$

$$(X,*) \longrightarrow (Z_f,*') \longrightarrow (C_f,*'),$$

CX being the reduced cone on X and Z_f the reduced mapping cylinder of f.

Remark 2.4: It is worth mentioning, in conclusion that a theorem similar to Theorem 2.1 holds with respect to a finite cyclic group of coefficients. Let $m \geq 1$ be an integer. Let $AG(X;m)$ be the quotient group $AG(X)/mAG(X)$ endowed with the quotient topology. By an argument strictly analogous to that outlined above, one shows that the groups $\pi_*(AG(X;m))$ and $H_*(X;\mathbb{Z}/m\mathbb{Z})$ are naturally isomorphic on the category of pointed separable polyhedra. This leads to a neat construction of Eilenberg-MacLane spaces. The zero component of $AG(S^n)$ is a $K(\mathbb{Z},n)$. The zero component of $AG(S^n;m)$ is a $K(\mathbb{Z}/m\mathbb{Z},n)$. By taking products, using the Structure Theorem, one thus obtains $K(G,n)$ spaces for any finitely generated abelian group G. It is shown in [2] that these spaces are CW complexes. Thus, if G is a CW topological abelian group whose homotopy groups are finitely generated,

$\bigoplus_{i=0}^{\infty} K(\pi_i(G), i + 1)$ is a classifying space for G (see Theorem 7.1 of [2]).

§3. Puppe's Proof using Semi-simplicial Methods

The reader unfamiliar with semi-simplicial topology can safely skip this section. A general reference is [9]. References here are all to Puppe's paper [11]. An alternative proof of Theorem 2.1 is sketched.

Whenever K is a simplicial set, let its set of q-simplexes be denoted by K_q, and if $x \in K_q$ let the ith face and i^{th} degeneracy of x be denoted respectively by $d_i x$ and $s_i x$. Let GK be the free simplicial abelian group generated by K: $(GK)_q$ is the free abelian group generated by K_q with the obvious induced face and degeneracy homomorphisms. There is an associated chain complex whose boundary operator is

$$d = \sum_{i=0}^{q} (-1)^i d_i : (GK)_q \longrightarrow (GK)_{q-1}$$

Let $H_q(GK) \equiv (GK)_q \cap \ker d / dGK_{q+1}$ be the qth homology group of this chain complex. If in particular K_q is the set of ordered (q + 1) - tuples of vertices which span some simplex of a given abstract simplicial complex $\mathcal{K}$, then $\mathcal{H}_q(G\mathcal{K})$ is the usual ordered simplicial homology group $H_q(\mathcal{K};\mathbb{Z})$.

Whenever G is a simplicial group, let $G_q^r = G_q \cap \left(\bigcap_{i \leq r} \ker d_i \right)$, $0 \leq r \leq q$. The qth simplicial homotopy group of G is defined to be $\pi_q(G) \equiv G_q^q / d_{q+1}(G_{q+1}^q)$.

If K is a simplicial pointed set with base point $* \in K_0$, then $*$ generates a subcomplex L of K consisting of a "degenerate base point" in each K_q. Let $\Gamma K \equiv GK/GL$ be the quotient simplicial abelian group. $\mathcal{H}_q(\Gamma K)$ and $\pi_q(\Gamma K)$ are defined as before. The following theorems of simplicial homotopy theory are fundamental.

<u>Theorem 3.1</u>: (Moore, see e.g. Theorem 2 of [11]) <u>If K is a simplicial pointed set, then there is a natural isomorphism between $\mathcal{H}_*(\Gamma K)$ and $\pi_*(\Gamma K)$.</u>

<u>Theorem 3.2</u>: (see e.g. Theorem 1 of [11]) <u>If G is a simplicial group, there is a natural isomorphism between $\pi_*(G)$ and $\pi_*(|G|,*)$</u>

<u>where</u> $|G|$ <u>denotes the geometric realization of</u> G, <u>and</u> $* \in G_0$ <u>is the identity element of</u> G_0.

With this preparation one is ready to start. Let $(X,*)$ be a pointed separable polyhedron, triangulated by the countable abstract simplicial complex $\mathcal{K}$. Let K be the countable simplicial set associated with $\mathcal{K}$ as above. The above discussion together with 3.1 and 3.2 yields natural isomorphisms

$$\tilde{H}_*(X) \cong H_*(X,\{*\}) \cong H_*(\mathcal{K},\{*\}) \cong \mathcal{H}_*(\Gamma K) \cong \pi_*(\Gamma K) \cong \pi_*(|\Gamma K|,*) .$$

On the other hand, the functorial properties of AG (see §2) yield a natural isomorphism

$$\pi_*(AG(X,*)) \cong \pi_*(AG(|K|,*)) .$$

In view of Remark 2.2, the Dold-Thom Theorem 2.1 will have been reproved if a natural weak homotopy equivalence can be established between $AG(|K|,*)$ and $|\Gamma K|$. In fact Puppe establishes a natural homeomorphism between them (see §2.10 of [11] and in particular 2.10(12)).

The idea of his proof is easily understood. It rests principally on the fact that the geometric realization of a cartesian product of two (countable) simplicial sets is naturally homeomorphic to the cartesian product of their geometric realizations. One need only review the passage $(|K|,*)$ to $SP^q(|K|)$ to $SP(|K|,*)$ to $AG(|K|,*)$ given in §2, and at each stage "do the same thing" in the category of pointed simplicial sets: given $(K,*)$ define, by analogy, $SP^q(K)$, $SP(K,*)$ and $AG(K,*)$. At each stage there will be a natural homeomorphism from the topological space to the geometric realization of its simplicial analogue. The resulting simplicial abelian group $AG(K,*)$ will be precisely ΓK.

For similar reasons, one obtains a new proof of the mod m version of the Dold-Thom Theorem (see 2.4).

§4. The Geometrical Approach

The main theorem is 4.2. The necessary definitions are given, together with some heuristic discussion. The proof is sketched. Its relationship to Almgren's Theorem is discussed in Remark 4.3, and to the Dold-Thom Theorem in Remark 4.4. Possible extensions are mentioned in Remark 4.6 .

Let X be a metric space (with a chosen metric). Let $Z_0(X)$ be the abelian group of reduced singular 0-cycles in X with integer coefficients i.e., the subgroup of the free abelian group generated by X consisting of those elements whose coefficients sum to zero. We will not regard X as pointed but we remark that if $* \in X$, there is an isomorphism of abelian groups between $AG(X,*)$ and $Z_0(X)$ which maps $\sum_i n_i x_i \in AG(X,*)$ to $\sum_i n_i x_i - (\sum_i n_i)* \in Z_0(X)$. But while $AG(X,*)$ carries a "weak" topology, we will give $Z_0(X)$ a metric topology as follows.

We will be concerned with $Z_0(X)$ and with the abelian group $\Delta_1(X)$ of singular 1-chains in X with integer coefficients. In either case let us say that a chain $\sum_i n_i \sigma_i$ is in <u>lowest terms</u> if each σ_i is a generator and $\sigma_i = \sigma_j$ if and only if $i = j$. If $z \in Z_0(X)$ and $z = \sum_i n_i x_i$ in lowest terms, the <u>mass</u> of z is $M(z) = \sum_i |n_i|$. $M(0) = 0$. If $c \in \Delta_1(X)$ and $c = \sum_i n_i \sigma_i$ in lowest terms, the <u>mass</u> of c is $M(c) = \sum_i |n_i| \operatorname{diam}(\sigma_i)$ where $\operatorname{diam}(\sigma_i)$ is the diameter in X of the image, $\sigma_i(\Delta^1)$, of the singular 1-simplex $\sigma_i : \Delta^1 \longrightarrow X$.

Again $M(0) = 0$. The Whitney metric on $Z_0(X)$ is

$$\rho(z_1,z_2) = \inf \{M(z_1 - z_2 + \partial c) + M(c) | c \in \Delta_1(X)\}.$$

The name is explained in §5, as is the motivation for considering such a metric. Here we simply observe (i) ρ really is a metric; (ii) ρ is translation invariant and makes $Z_0(X)$ into a topological group; (iii) for nearby chains ρ becomes simpler —— if $\rho(z_1,z_2) < 1$ then

$$\rho(z_1,z_2) = \inf \{M(c) | c \in \Delta_1(X) \text{ and } \partial c = z_1 - z_2\};$$

It is important that the reader gain some feel for this metric. For example, if x' and x'' are nearby points in X, and if $c = \sum_i n_i x_i \in Z_0(X)$ is in lowest terms then c is near $c + x' - x''$ even if all the x_i's are very far away from x' and x''. The best physical analogy is obtained by thinking of c as a system of particles and anti-particles, $n_i x_i$ standing for $|n_i|$ particles [resp. anti-particles] at position x_i if n_i is positive [resp. negative]. Then $c + x' - x''$ differs from c by a particle/anti-particle pair "spontaneously" created far away. This is a familiar idea in modern physics. In fact, by introducing more complicated coefficient groups, as we do in 4.6, this analogy could be pushed quite far (and far away from our topic, to which we now return).

The metric ρ depends on the metric of X because the mass of a 1-chain depends on diameters. It can happen that topologically equivalent metrics on X induce inequivalent metrics on $Z_0(X)$. In fact this can happen even when X is the closed unit interval. Furthermore a (continuous) map between two metric spaces need not induce a continuous homomorphism between the 0-cycles. These apparent difficulties are explained by

Proposition 4.1: *Z_0 is a functor from the category of metric spaces and Lipschitz maps to the category of topological abelian groups and continuous homomorphisms.*

Because of 4.1 let us recall that every C^1-map between compact differentiable manifolds, and every piecewise linear map between compact polyhedra is Lipschitz (when the spaces involved are given reasonable metrics).

We now construct, for each metric space X, a chain complex $\{\Gamma_*(X), \partial\}$ whose homology groups will be the homotopy groups of $Z_0(X)$ based at 0. The notation Γ has been chosen to emphasize the similarity with the simplicial construction given in §3. To establish our standard simplexes, let us choose once and for all a sequence $\{0,1,2,\cdots\}$ of linearly independent points in Hilbert Space. Let Δ^q be the convex hull of the vertices $\{0,1,\cdots,q\}$. Thus Δ^{q-1} lies in Δ^q as the face opposite q. Let $\Gamma_q(X)$ be the abelian group of all (continuous) maps $f:\Delta^q \longrightarrow Z_0(X)$ such that f maps every face of Δ^q except perhaps the last face, Δ^{q-1}, to $0 \in Z_0(X)$. Define $\partial:\Gamma_q(X) \longrightarrow \Gamma_{q-1}(X)$ by $\partial(f) = f|\Delta^{q-1}$ if $q > 0$. Define $\Gamma_q(X) = 0$ if $q < 0$. Clearly we have a chain complex with the required properties. In fact the construction of Γ_*, when extended in the obvious manner to Lipschitz maps, makes it a functor.

Next we construct a natural chain map from $\Delta_*(X)$ to $\Gamma_*(X)$, where $\Delta_*(X)$ is the augmented singular chain complex of X (for convenience we take $\Delta_0(X) \equiv Z_0(X)$ and $\Delta_{-1}(X) = \{0\}$: this represents a change from the notation of §1, but it will not cause trouble). The chain map is to associate with a singular q-chain c the map $\hat{c} \in \Gamma_q(X)$ defined by

$$\hat{c}(t) = \sum_i n_i \sigma_i(t) ,$$

where $t \in \Delta^q$, $\sum_i n_i \sigma_i$ is the first barycentric subdivision†, sd(c), of the chain c, and addition takes place in $Z_0(X)$. The case $q = 2$ and $c = \tau_1 + \tau_2$ (the sum of two singular 2-simplexes) is illustrated in Figure 1.

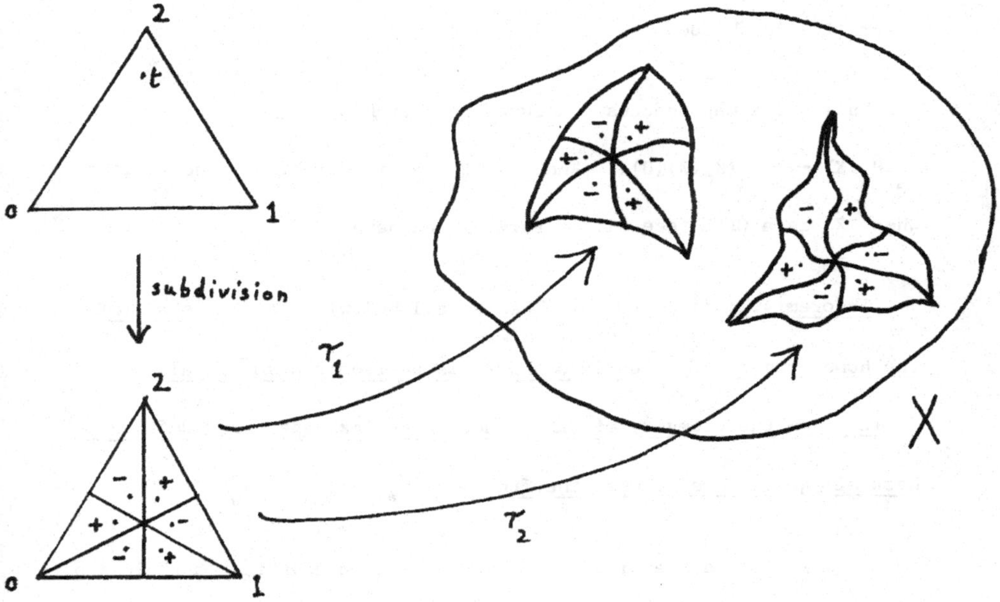

Figure 1

†Our definition of subdivision of singular chains has to be technically different from Spanier's on page 177 of [12]. Ours is obtained from his by redefining $\beta_n(x_0,\cdots,x_q) = (x_0,\cdots,x_q,b_n)$.

While we omit proofs here, the reader will gain all the necessary insight by verifying for the case illustrated that i) if $c \in \Delta_q(X)$ then $\widehat{(-c)}(t) = -\hat{c}(t)$; ii) if $c = c_1 + c_2$ then $\hat{c}(t) = \hat{c}_1(t) + \hat{c}_2(t)$; iii) if t is on any face of Δ^q except the last, $\hat{c}(t) = 0$; iv) if t is on the last face (i.e., $t \in \Delta^{q-1}$) $\hat{c}(t) = \widehat{\partial c}(t)$; v) $\hat{c}$ is continuous. In short, the map $c \longmapsto \hat{c}$ is a chain map $\Delta_q(X) \longrightarrow \Gamma_q(X)$. Naturality is obvious.

This chain map induces a natural homomorphism $F_q : \tilde{H}_q(X) \longrightarrow \pi_q(Z_0(X), 0)$. When is F_q an isomorphism? The answer: when X is a CLNR (see §1). Formally we have

<u>Theorem 4.2</u>: ([7]; compare Theorem 6.1 below) *If X is a CLNR, the homomorphism F_q, defined above, is an isomorphism for all $q \geq 0$. In fact, on the category of CLNR's and Lipschitz maps, F_* defines a natural equivalence between the functors $\tilde{H}_*$ and $\pi_* \circ Z_0$.*

We now discuss the proof. One would like to construct a chain map: $\Gamma_q(X) \longrightarrow \Delta_q(X)$ which is a chain homotopy inverse to the chain map already described. But $\Gamma_q(X)$ is not (in any obvious sense) free, so no inverse suggests itself on the level of chain complexes. We pass to the level of groups, first sketching the proof that F_q is a monomorphism.

Let $z \in Z_q(X)$ be such that $\hat{z}:\Delta^q \longrightarrow Z_0(X)$ represents the trivial element of $\pi_q(Z_0(X),0)$. Then there exists $f:\Delta^{q+1} \longrightarrow Z_0(X)$ lying in $\Gamma_{q+1}(X)$ such that $\partial f \equiv f|\Delta^q = \hat{z}$. We are to construct $c \in \Delta_{q+1}(X)$ such that ∂c is homologous to z. Let $sd^n\delta^{q+1}$ be its nth barycentric subdivision and let $\Delta_*(sd^n\delta^{q+1})$ be the (unaugmented!) ordered chain complex of $sd^n\delta^{q+1}$ (see Spanier [12]). Since X is an ANR there is a retraction $r:U \longrightarrow X$ where U is a locally convex neighborhood. For suitably large n one constructs (non-canonical) chain maps $\phi_*:\Delta_*(sd^n\delta^{q+1}) \longrightarrow \Delta_*(U)$ such that $\phi_0(v) = f(v)$ for each vertex v, and each "connected component" of $\phi_r(\sigma)$, $r \geq 1$, has small diameter in U. The compactness of X and the local convexity of U are used here. Let the image under $r_* \circ \phi_q$ of the fundamental class of $sd^n\delta^{q+1}$ be $c \in \Delta_{q+1}(X)$. If n and ϕ_* are suitably chosen, ∂c will be $sd^{n+1}z$, which is homologous to z. The essential ingredient can be described heuristically as "simplicial approximation" of continuous maps $\Delta^q \longrightarrow Z_0(X)$ by chain maps $\Delta_*(sd^n\delta^q) \longrightarrow \Delta_*(X)$. One uses only the fact that X is a compact ANR.

By contrast, the essential ingredient of the proof that F_q is onto can be described as "replacement" of chain maps $\Delta_*(sd^n\delta^q) \longrightarrow \Delta_*(X)$ by continuous maps $\Delta^q \longrightarrow Z_0(X)$. Let $f:\Delta^q \longrightarrow Z_0(X)$ be continuous, and let $\partial f = 0$. Then $f(\dot{\Delta}^q) = 0$. We are to construct $z \in Z_q(X)$ such that $\hat{z}$ is homotopic to f leaving $\dot{\Delta}^q$ fixed. Let K be a locally finite complex triangulating $\Delta^q \times [0,1)$ so that on the level $\Delta^q \times \{n/n+1\}$, K agrees with $sd^n\delta^q$, Let $\dot{K}$ triangulate $\dot{\Delta}^q \times [0,1)$. By a method similar to that used in the monomorphism case, we can construct a chain map $\Phi:\Delta_*(K) \longrightarrow \Delta_*(U)$ such that $\Phi = 0$ on $\Delta_*(\dot{K})$ and, for suitably large n, Φ agrees with f on the vertices of $sd^n\Delta^q \times \{n/n+1\}$. Let $z_0 \in Z_q(U)$ be the Φ-image of the fundamental class at level 0 (i.e., $z_0 = \Phi_q(\delta^q \times \{0\})$). Our method of choosing Φ and our method of replacing chain maps by continuous maps (too technical to outline here) give us a uniformly continuous homotopy $H:\Delta^q \times [0,1) \longrightarrow Z_0(U)$ such that $H|\Delta^q \times \{0\} = \hat{z}_0$ and $H|\Delta^q \times \{n/n+1\}$ agrees with Φ, hence also with f, on the vertices of $sd^n\delta^q$ (n large). Thus H extends to $\Delta^q \times [0,1]$, agreeing with f on level 1. The retraction

$r: U \longrightarrow X$ is Lipschitz. Hence $r_{\#} \circ H$ is a homotopy in $Z_0(X)$ between f and $r_{\#} \circ \hat{z}_0$. But $r_{\#} \circ \hat{z}_0 = \widehat{r_*(z_0)}$, so the required z is $r_*(z_0)$.

This completes our discussion of the proof itself but some comments are still needed.

Remark 4.3: Theorem 4.2 is a generalization of a special case of Almgren's Theorem [1] described in §6 below. Almgren uses a metric different from, but topologically equivalent to ρ , and he uses integral current homology rather than singular homology. However all of the ideas sketched above have analogues in his work. Almgren's Theorem is a topological theorem with an analytic proof. We have separated the topology from the analysis. This is not to find fault with [1]. There the purposes and the context were different, as will be described in the following sections. Our claim is, rather, that our proof, being totally topological while at the same time using some important ideas of geometric integration theory, serves as a bridge on the road from "pure" homology to geometric integration theory.

Remark 4.4: Theorem 4.2 reproves the Dold-Thom Theorem 2.1 in many cases. We remarked at the start of this section that $AG(X,*)$ and

$Z_0(X)$ are isomorphic as topological groups. In fact this isomorphism gives a natural weak homotopy equivalence $AG(X,*) \longrightarrow Z_0(X)$ when X is a CLNR: the proof of 4.2 gives this easily. The compactness of X seems to be essential when dealing with $Z_0(X)$.

Remark 4.5: If X is a finite dimensional CLNR then $Z_0(X)$ is an ANR. If X is an infinite-dimensional CLNR, we do not know how to prove that $Z_0(X)$ is an ANR, but we can prove that $Z_0(X)$ has the homotopy type of a CW complex (a property which all ANR's possess). See [8] for more details.

Remark 4.6: Throughout we have dealt with integer coefficients. Dold and Thom also prove their theorem with a finite cyclic group of coefficients (see Remark 2.4). If we identify $AG(X,*)$ with $Z_0(X)$ and $AG(X,*;m)$ with $Z_0(X;\mathbb{Z}/m\mathbb{Z})$ as above, we can say that Dold and Thom converted homology with integer - or finite-cyclic coefficients into homotopy of 0 - chains with integer or finite-cyclic coefficients. It is trivial then to extend their result to finitely generated (abelian) coefficient groups, as explained in 2.4. But we can do better. Our method permits a Dold-Thom Theorem for CLNR's with arbitrary (abelian) coefficient group G. We simply give G a translation invariant metric inducing the discrete topology: we denote the distance from $g \in G$ to $0 \in G$ by $|g|$ and proceed exactly as we did with $\mathbb{Z}$ coefficients. We remark that even when $G = \mathbb{Z}$, this allows us more than one possible notion of mass: the geometrical consequences are strange. We also remark that with the discrete metric on $\mathbb{R}$, $Z_0(X;\mathbb{R})$ does not become a topological vector space (scalar multiplication is not continuous) which is just as well, since we claim that $\pi_*(Z_0(X;\mathbb{R}))$ is isomorphic to $H_*(X;\mathbb{R})$ when X is a CLNR.

§5. Whitney's Flat Norm

Polyhedral chains are defined here, and the geometrical motivation for using metrics such as the metric ρ defined in §4 is given. This section is intended as an introduction to §6. The source is [13].

A (non-degenerate affine) q-simplex ($q \geq 0$) in $\mathbb{R}^n$ is the convex hull of $(q+1)$ affinely independent points of $\mathbb{R}^n$. Let $S_q(\mathbb{R}^n)$ be the real vector space generated by the oriented q-simplexes modulo the equivalence relation $\sigma_1 + \sigma_2 = 0$ if σ_1 and σ_2 differ only in their orientations. Define $S_q(\mathbb{R}^n) = 0$ if $q < 0$. The boundary $\partial : S_q(\mathbb{R}^n) \longrightarrow S_{q-1}(\mathbb{R}^n)$ is defined in the usual manner.

Now let K be a simplicial complex subdividing some q-simplex σ (so that $|K| = \sigma$) and let the q-simplexes of K be $\tau_1, \cdots, \tau_m$. Choose an orientation for σ and, abusing notation, let this oriented simplex be represented in $S_q(\mathbb{R}^n)$ as $\varepsilon\sigma$ where ε is $+1$ or -1 depending on the orientation chosen. Let each τ_i be compatibly oriented and represented in $S_q(\mathbb{R}^n)$ as $\varepsilon_i\tau_i$ where ε_i is $+1$ or -1. Then the chain $\sum_i \varepsilon_i\tau_i$ is the <u>subdivision</u> of $\varepsilon\sigma$ induced by K. If $c = \sum_j a_j\sigma_j$ is a chain in $S_q(\mathbb{R}^n)$ with each $a_j \in \mathbb{R}$, and if $\sum_i \varepsilon_{ij}\tau_{ij}$ is a subdivision of $(+1)\sigma_j$, then the chain $\sum_{i,j} a_j\varepsilon_{ij}\tau_{ij}$ is called a <u>subdivision</u> of c. Two chains of $S_q(\mathbb{R}^n)$ are <u>equivalent</u> if they have a common subdivision. This is clearly an equivalence relation compatible with the boundary homomorphism; hence there is a quotient chain complex $\{P_q(\mathbb{R}^n)\}$. The elements of $P_q(\mathbb{R}^n)$ are called (real) <u>polyhedral</u> q-<u>chains</u>. Note that $P_q(\mathbb{R}^n)$ is still a vector space but with no obvious basis: we will return to this point in §6.

The corresponding cochains and coboundary are denoted by

$\delta: P^q(\mathbb{R}^n) \longrightarrow P^{q+1}(\mathbb{R}^n)$.

One is guided by thinking of polyhedral q - chains as "domains of q - dimensional integration," of q - cochains as "q - dimensional integrands" and (given $c \in P_q$ and $X \in P^q$) of the number $X(c)$ as "the integral of X over c," $\int_c X$. Two basic properties of integrals hold, namely that $\int_{c_1+c_2} X = \int_{c_1} X + \int_{c_2} X$, and, if $X \in P^q(\mathbb{R}^n)$ and $e \in P_{q+1}(\mathbb{R}^n)$, $\int_{\partial e} X = \int_e \delta X$ ("Stokes' Theorem"). This fundamental analogy between (real) chains and cochains on the one hand, and domains and integrands on the other, suggests that one should confine attention to cochains which enjoy two special "continuity" properties:

Property 1: For a given q - cochain X there should be a number N_1 such that $|X(\sigma)| \leq N_1|\sigma|$ whenever $\sigma(\equiv (+1)\sigma)$ is an oriented q - simplex with q - dimensional volume $|\sigma|$, defined as in linear algebra.

Property 2: For a given q - cochain X, there should be a number N_2 such that $|X(\partial\sigma)| \leq N_2|\sigma|$ whenever σ is an oriented $(q+1)$ - simplex with $(q+1)$ - dimensional volume $|\sigma|$.

Whitney calls cochains having Properties 1 and 2 <u>flat cochains</u>. The smallest possible N_1 in Property 1 is

$$|X| = \sup \left\{ \frac{|X(\sigma)|}{|\sigma|} \,\middle|\, \sigma \text{ is an oriented } q\text{-simplex} \right\}.$$

The smallest N_2 is

$$|\delta X| = \sup \left\{ \frac{|X(\partial\sigma)|}{|\sigma|} \,\middle|\, \sigma \text{ is an oriented } (q+1)\text{-simplex} \right\}.$$

If X is a flat q - cochain its <u>flat norm</u> is

$$|X|^\flat = \max\{|X|, \ |\delta X|\}$$

(it makes the vector space of flat q - cochains a Banach space). But our real interest is in the dual norm on the vector space $P_q(\mathbb{R}^n)$, namely

$$|c|^\flat = \inf\{N \,\big|\, |X(c)| \le |X|^\flat . N \quad \text{for all flat } q\text{-cochains } X\}.$$

$|\cdot|^\flat$ is called the <u>flat norm</u> on $P_q(\mathbb{R}^n)$. As Whitney points out

(p. 154 of [13]), this flat norm has a beautiful geometrical interpretation as follows. Let $c \in P_q(\mathbb{R}^n)$. Write $c = \sum_i a_i \sigma_i$ where the σ_i's are non-overlapping q-simplexes. Define the mass of c to be $M(c) = \sum_i |a_i| |\sigma_i|$ (where $|\sigma_i|$ is the q-dimensional volume of σ_i as before). $M(c)$ is clearly independent of the simplex decomposition $\sum_i a_i \sigma_i$. Then the flat norm becomes

$$|c|^\flat = \inf \{M(c - \partial e) + M(e) \mid e \in P_{q+1}(\mathbb{R}^n)\}.$$

Thus if $c_1, c_2 \in P_q(\mathbb{R}^n)$, the distance between them is

$$|c_1 - c_2|^\flat = \inf \{M(c_1 - c_2 - \partial e) + M(e) \mid e \in P_{q+1}(\mathbb{R}^n)\}$$

a formula strongly reminiscent of the metric ρ used in §4. Distance is measured "homologically" in terms of the $(q+1)$-dimensional mass of "spanning chains" plus the q-dimensional mass of "unwanted pieces" of their boundaries.

§6. Integral currents and Almgren's Theorem

There are two ways of introducing integral currents. One way is to view them as those linear functionals on differential forms which can be thought of as chains with integer coefficients, and this is the natural approach if one is working with them (see [4] or [3] for this approach). The other way is shorter; we follow Fleming [6].

Start with $P_q(\mathbb{R}^n)$ as in §5. Look at the subgroup $P_q(\mathbb{R}^n;\mathbb{Z})$ of those polyhedral chains which have integer coefficients. Give $P_q(\mathbb{R}^n;\mathbb{Z})$ the metric

$$d(c_1,c_2) = \inf\left\{M(c_1 - c_2 - \partial e) + M(e)\,\middle|\, e \in P_{q+1}(\mathbb{R}^n;\mathbb{Z})\right\}$$

(Mass of chains in $P_q(\mathbb{R}^n)$ was defined in §5.) Let the completion of $P_q(\mathbb{R}^n;\mathbb{Z})$ with respect to d be $F_q(\mathbb{R}^n;\mathbb{Z})$, abbreviated to F_q. Extend the notion of mass to F_q as follows: if $c \in F_q$, $M(c)$ is the smallest number $\lambda \leq \infty$ such that there is a sequence $\{c_j\}$ in $P_q(\mathbb{R}^n;\mathbb{Z})$ converging to c, with $M(c_j)$ converging to λ. Extend the boundary homomorphism ∂ to F_q in the obvious way, $\partial: F_q \longrightarrow F_{q-1}$. The elements of F_q are called <u>flat chains</u>. A flat chain c is <u>supported</u> by a compact subset X of $\mathbb{R}^n$ if given any neighborhood U of X there is a sequence $\{c_j\}$ in $P_q(\mathbb{R}^n;\mathbb{Z})$

converging to c, such that each c_j lies in U. The smallest such compact set X is called the support of c. The abelian group of q-dimensional integral currents in X is

$$I_q(X) = \{c \in F_q \mid M(c) < \infty, \ M(\partial c) < \infty \text{ and } c \text{ is supported by } X\}.$$

The metric, when restricted to $I_q(X)$ is equivalent to

$$d_X(c_1,c_2) = \inf \{M(c_1 - c_2 - \partial e) + M(e) \mid e \in I_{q+1}(X)\}$$

(see 3.1 of [6] and 1.19 of [1]). $\partial : I_q(X) \longrightarrow I_{q-1}(X)$ is continuous. If X is a CLNR (in $\mathbb{R}^n$) the resulting integral current homology groups of X, $\mathcal{H}_*(X)$, are isomorphic to the singular homology groups of X with integer coefficients (see Theorem 5.11 of [4]). Let the integral cycle groups be

$$Z_q(X) = \{z \in I_q(X) \mid \partial z = 0\} \quad \text{if} \quad q \geq 1$$

$$Z_0(X) = \{z \in I_0(X) \mid \text{the coefficient sum of } z \text{ is } 0\}.$$

The definition of $Z_0(X)$ makes sense because all 0-dimensional integral currents in X are polyhedral (when X is compact). Thus $Z_0(X)$ as defined here is exactly the same as $Z_0(X)$ defined in §4. The metrics ρ and d_X are equivalent.

Theorem 6.1: (Almgren: Theorem 7.5 of [1]) *Let X be a CLNR lying in $\mathbb{R}^n$. Then for each $m \geq 0$ and $q \geq 0$ the (reduced) integral current homology group $\tilde{H}_{q+m}(X)$ is isomorphic to $\pi_q(Z_m(X),0)$, In fact, on the category of CLNR's and Lipschitz maps there is a natural equivalence between the functors $\tilde{H}_{q+m}$ and $\pi_q \circ Z_m$.*

One can easily indicate how to define the homomorphism $\tilde{H}_{q+m}(X) \longrightarrow \pi_q(Z_m(X),0)$ in a simple case. Let X be a compact sub-polyhedron of the unit ball $B^n \subset \mathbb{R}^n$. Any homology class in $\tilde{H}_{q+m}(X)$ may be represented by a polyhedral cycle z with integer coefficients (see Theorem 5.5 of [4]): let $z = \sum_i a_i\sigma_i$. Assume $q + m \leq n$. Let $\{P_t \mid -\infty < t < +\infty\}$ be a family of parallel hyperplanes in $\mathbb{R}^n$, where the perpendicular distance from P_t to the origin in $\mathbb{R}^n$ is t. Choose (compatible) orientations for the P_t's and let U_t be the closure of the set of points on the positive side of P_t. Choose the direction of the P_t's so that no positive-dimensional face of any σ_i lies in any P_t (general position!). Associate with z the loop of $(q+m-1)$-dimensional polyhedral cycles whose t-point $(-1 \leq t \leq 1)$ is $\partial(z \cap U_t)$. Repeat this process q times using q different families of hyperplanes, all in general position with respect to z in the sense described. In this way one associates with z a representative of an element of $\pi_q(Z_m(X),0)$. This, when done on each homology class, gives the required homomorphism.

Even when $m = 0$ this homomorphism cannot be defined on the level of chains, from $I_q(X)$ to $\Gamma_q(X)$. The "slicing" of a cycle had to be by hyperplanes in general position with respect to that cycle.

Almgren's proof that this homomorphism is an isomorphism is very difficult. In the case $m = 0$, the monomorphism part is not unlike the monomorphism part of the proof of Theorem 4.2. But the epimorphism part is much harder.

§7. Parametrized Surfaces

$\Gamma_q(X)$ is the set of continuous maps from Δ^q to $Z_0(X)$ which map all faces to 0 except perhaps the last, Δ^{q-1} (see §4). The elements of $\Gamma_q(X)$ can be thought of as parametrized q-dimensional surfaces in X of varying topological type, in much the same sense as the maps from some fixed q-manifold M into X can be thought of as parametrized q-dimensional surfaces in X of the (singular!) topological type of M. Now, neither part of the last sentence is to be taken too literally, but as a philosophical statement it is correct. Here are some reasons.

An integral polyhedral q-chain lying in X, when written in the form $\sum_i a_i\sigma_i$ where the σ_i are non-overlapping, can be thought of as a finite union of oriented q-simplexes, the simplex σ_i being oriented by the sign of a_i, and being counted with multiplicity $|a_i|$. If this is considered to be a q-dimensional "surface," then there are many ways of associating elements of $\Gamma_q(X)$ with it. One may slice it with hyperplanes, as described at the end of §6; or one may regard it as a singular q-chain, $c \in \Delta_q(X)$, and obtain $\hat{c} \in \Gamma_q(X)$, as described in §4. Either way, the "parameterizing" map $f:\Delta^q \longrightarrow Z_0(X)$ has the following property: if a point x appears in $\sigma_{i_1} \cap \cdots \cap \sigma_{i_m}$ and in no other σ_i, and if $n_t \in \mathbb{Z}$ is the coefficient of x in $f(t)$, where $t \in \Delta^q$, then n_t is non-zero for only finitely many values of t, and $\sum_{j=1}^{m} a_{i_j} = \sum_{t \in \Delta^q} n_t$. (Parenthetically we remind the reader that $m = 1$ if x lies in the interior of any σ_i, since the σ_i's are non-overlapping.)

Now, if one replaces the σ_i's by differentiable submanifolds M_i^q of $\mathbb{R}^n$ which lie in X, the formal sum $\sum_i a_i M_i$ can still be regarded as an integral current, and it is again possible to parametrize it with elements of $\Gamma_q(X)$ as in the polyhedral case. What is more, if M is a q-dimensional submanifold of $\mathbb{R}^n$ lying in X, and if M' is obtained from M by removing a small q-ball, then M and M' can be parametrized by maps from Δ^q to $Z_0(X)$ which are uniformly close. This is the sense in which in $\Gamma_q(X)$ one may pass continuously from one topological type to another.

Not every integral current has a manifold or polyhedron as its support. In the same way, not every element of $\Gamma_q(X)$ can reasonably be considered to have a q-manifold or polyhedron as its support. But then, not every continuous map from Δ^q into X can be regarded as a singular q-manifold: Peano showed this! However the pathology of the support is slight if the integral current is "taut": the justification for this last claim lies in the following Federer-Fleming solution to (one of the versions of) Plateau's problem. $c \in I_q(\mathbb{R}^n)$ is a <u>current of least mass</u> if $M(c) \leq M(c')$ for all $c' \in I_q(\mathbb{R}^n)$ such that $\partial c' = \partial c$. c is <u>minimal</u> if, locally, c is of least mass.

<u>Theorem 7.1</u>: (see Theorem 8.13 [4] and Theorems 1 and 2 of [5].) <u>If</u> $z \in Z_{q-1}(\mathbb{R}^n)$, <u>then there exist minimal integral currents</u> $c \in I_q(\mathbb{R}^n)$ <u>such that</u> $\partial c = z$. <u>If</u> M <u>is the support of such a minimal current</u> c, <u>and if</u> N <u>is the support of</u> ∂c, <u>then</u> $M \setminus N$ <u>is</u>

locally homeomorphic to $\mathbb{R}^q$, except perhaps on a closed set S of zero q-dimensional Hausdorff measure. If $n = 3$ and $q = 2$, S is empty.

This is one of the key theorems concerning integral currents. The method of proof is measure-theoretic. It would be possible to define a notion of mass directly on $\Gamma_q(\mathbb{R}^n)$, and to restate this theorem in a parametrized form. However the exercise would be phony and superficial unless one could devise a new proof using little or no measure theory.

§8. Concluding Remarks and Conjectures

Having traced the Dold-Thom construction and its variants from algebraic topology to the theory of integration, it seems sensible to conclude with some personal remarks on why I have written this paper. In general terms, I am interested in connections between analysis and topology. More specifically I approach this work with a background in infinite-dimensional topology, which is, roughly, the study of suitable infinite-dimensional topological spaces as manifolds. I conjecture that the spaces of integral currents discussed here are infinite-dimensional manifolds, and since the spaces are incomplete, they can only be locally homeomorphic to incomplete models. To be more specific, if ℓ_2 denotes the usual separable Hilbert space of square-summable sequences, let ℓ_2^f be the dense, incomplete, linear subspace consisting of those sequences which have only finitely many non-zero entries. I conjecture that _if_ X _is a finite-dimensional_ CLNR _which is not discrete then_ $Z_0(X)$ _is locally homeomorphic to_ ℓ_2^f. In support of this conjecture I make two remarks. Firstly, the fact that there are no "obvious" local ℓ_2^f charts in $Z_0(X)$ is neither here nor there: the space of piecewise linear maps from a compact non-discrete polyhedron to itself (with the topology of uniform convergence) is locally homeomorphic to ℓ_2^f, and there are no obvious ℓ_2^f charts in that case either - especially when the polyhedron is not a manifold [14]. Secondly, using theorems of Haver [15] and Torunczyk [16] I can state as a fact that if X is as above then $Z_0(X) \times \ell_2^f$ is locally homeomorphic to ℓ_2^f. The conjecture therefore reduces to the problem of "absorbing" the ℓ_2^f factor: this is a typical problem in infinite-dimensional topology. [I can also guess what should be the local model for $Z_q(X)$ when $q > 0$. But since I cannot yet show that these spaces are ANR's it would be premature to state my conjecture.]

Why should one care? I care because, on the one hand, infinite-dimensional topology gives considerable insight into the geometry of such manifolds; and, on the other hand, the subject will be enriched by a completely new source of naturally arising examples of such manifolds.

I will end with two remarks on the methods discussed in this paper (having no connection with infinite-dimensional topology). We have passed from algebraic topology, through geometric topology, to geometric integration theory. It is worth asking: where does the topology end and the analysis begin? and why can we not state a topological analogue of Almgren's Theorem for cycles of dimension > 0? It might be said that the analysis begins with the introduction of mass in dimension ≥ 2. Mass in dimension 0 is obtained arithmetically by counting, and mass in dimension 1 involves length which in its most primitive form is a metric, hence a topological, concept. This observation is the basis of the proof outlined in §4. But without a topological notion of mass in higher dimensions (e.g., the "mass" of a singular simplex) it does not seem possible to state, much less prove, a higher dimensional analogue of Almgren's Theorem 6.1 using singular chains.

It is also natural to ask if a proof similar to that of Dold and Thom still works for $Z_0(X)$. Specifically is there a version of Proposition 2.3 telling us that

$$Z_0(X) \longrightarrow Z_0(X/A)$$

is a principal fiber bundle with fiber $Z_0(A)$. I do not know. The problem seems to lie in the fact that $Z_0(A)$ is not complete.

Otherwise, a selection theorem due to Michael would give an affirmative answer.

References

1. F. J. Almgren, The homotopy groups of integral cycle groups, Topology 1 (1962), 257-299.

2. A. Dold and R. Thom, Quasifaserungen und Unendliche Symmetrische Produkte, Ann. Math. 67 (1958), 230-281.

3. H. Federer, Geometric Measure Theory, Springer Verlag, Berlin 1969.

4. H. Federer and W. Fleming, Normal and integral currents, Ann. Math. 72 (1960), 458-520.

5. W. Fleming, On the oriented Palteau problem, Rend. Circ. Math. di Palermo 11 (1962), 69-90.

6. __________, Flat chains over a finite coefficient group, Trans. Amer. Math. Soc. 121 (1966), 160-186.

7. R. Geoghegan, Singular chains, integral currents and the Dold-Thom construction (to appear).

8. ___________, A note on Lipschitz retracts (to appear).

9. J. P. May, Simplicial Objects in Algebraic Topology, Van Nostrand, Princeton, New Jersey, 1968.

10. M. C. McCord, Classifying spaces and infinite symmetric products, Trans. Amer. Math. Soc. 146 (1969), 273-298.

11. D. Puppe, Homotopie und Homologie in abelschen Gruppen- und Monoidkomplexen I, II, Math. Zeit., 68 (1958), 367-421.

12. E. Spanier, Algebraic Topology, McGraw-Hill Book Company, New York 1966.

13. H. Whitney, Geometric Integration Theory, Princeton University Press, Princeton, New Jersey, 1957.

14. R. Geoghegan, On spaces of homeomorphisms, embeddings, and functions, II: the piecewise linear case, Proc. Lond. Math. Soc., (3) 27, (1973), 463-483.

15. W. Haver, Locally contractible spaces that are absolute neighborhood retracts, Proc. Amer. Math. Soc. (to appear).

16. H. Torunczyk, Absolute retracts as factors of normed linear spaces, (preprint).

Vol. 310: B. Iversen, Generic Local Structure of the Morphisms in Commutative Algebra. IV, 108 pages. 1973. DM 16,-

Vol. 311: Conference on Commutative Algebra. Edited by J. W. Brewer and E. A. Rutter. VII, 251 pages. 1973. DM 22,-

Vol. 312: Symposium on Ordinary Differential Equations. Edited by W. A. Harris, Jr. and Y. Sibuya. VIII, 204 pages. 1973. DM 22,-

Vol. 313: K. Jörgens and J. Weidmann, Spectral Properties of Hamiltonian Operators. III, 140 pages. 1973. DM 16,-

Vol. 314: M. Deuring, Lectures on the Theory of Algebraic Functions of One Variable. VI, 151 pages. 1973. DM 16,-

Vol. 315: K. Bichteler, Integration Theory (with Special Attention to Vector Measures). VI, 357 pages. 1973. DM 26,-

Vol. 316: Symposium on Non-Well-Posed Problems and Logarithmic Convexity. Edited by R. J. Knops. V, 176 pages. 1973. DM 18,-

Vol. 317: Séminaire Bourbaki – vol. 1971/72. Exposés 400–417. IV, 361 pages. 1973. DM 26,-

Vol. 318: Recent Advances in Topological Dynamics. Edited by A. Beck, VIII, 285 pages. 1973. DM 24,-

Vol. 319: Conference on Group Theory. Edited by R. W. Gatterdam and K. W. Weston. V, 188 pages. 1973. DM 18,-

Vol. 320: Modular Functions of One Variable I. Edited by W. Kuyk. V, 195 pages. 1973. DM 18,-

Vol. 321: Séminaire de Probabilités VII. Edité par P. A. Meyer. VI, 322 pages. 1973. DM 26,-

Vol. 322: Nonlinear Problems in the Physical Sciences and Biology. Edited by I. Stakgold, D. D. Joseph and D. H. Sattinger. VIII, 357 pages. 1973. DM 26,-

Vol. 323: J. L. Lions, Perturbations Singulières dans les Problèmes aux Limites et en Contrôle Optimal. XII, 645 pages. 1973. DM 42,-

Vol. 324: K. Kreith, Oscillation Theory. VI, 109 pages. 1973. DM 16,-

Vol. 325: Ch.-Ch. Chou, La Transformation de Fourier Complexe et L'Equation de Convolution. IX, 137 pages. 1973. DM 16,-

Vol. 326: A. Robert, Elliptic Curves. VIII, 264 pages. 1973. DM 22,-

Vol. 327: E. Matlis, 1-Dimensional Cohen-Macaulay Rings. XII, 157 pages. 1973. DM 18,-

Vol. 328: J. R. Büchi and D. Siefkes, The Monadic Second Order Theory of All Countable Ordinals. VI, 217 pages. 1973. DM 20,-

Vol. 329: W. Trebels, Multipliers for (C, α)-Bounded Fourier Expansions in Banach Spaces and Approximation Theory. VII, 103 pages. 1973. DM 16,-

Vol. 330: Proceedings of the Second Japan-USSR Symposium on Probability Theory. Edited by G. Maruyama and Yu. V. Prokhorov. VI, 550 pages. 1973. DM 36,-

Vol. 331: Summer School on Topological Vector Spaces. Edited by L. Waelbroeck. VI, 226 pages. 1973. DM 20,-

Vol. 332: Séminaire Pierre Lelong (Analyse) Année 1971-1972. V, 131 pages. 1973. DM 16,-

Vol. 333: Numerische, insbesondere approximationstheoretische Behandlung von Funktionalgleichungen. Herausgegeben von R. Ansorge und W. Törnig. VI, 296 Seiten. 1973. DM 24,-

Vol. 334: F. Schweiger, The Metrical Theory of Jacobi-Perron Algorithm. V, 111 pages. 1973. DM 16,-

Vol. 335: H. Huck, R. Roitzsch, U. Simon, W. Vortisch, R. Walden, B. Wegner und W. Wendland, Beweismethoden der Differentialgeometrie im Großen. IX, 159 Seiten. 1973. DM 18,-

Vol. 336: L'Analyse Harmonique dans le Domaine Complexe. Edité par E. J. Akutowicz. VIII, 169 pages. 1973. DM 18,-

Vol. 337: Cambridge Summer School in Mathematical Logic. Edited by A. R. D. Mathias and H. Rogers. IX, 660 pages. 1973. DM 42,-

Vol: 338: J. Lindenstrauss and L. Tzafriri, Classical Banach Spaces. IX, 243 pages. 1973. DM 22,-

Vol. 339: G. Kempf, F. Knudsen, D. Mumford and B. Saint-Donat, Toroidal Embeddings I. VIII, 209 pages. 1973. DM 20,-

Vol. 340: Groupes de Monodromie en Géométrie Algébrique. (SGA 7 II). Par P. Deligne et N. Katz. X, 438 pages. 1973. DM 40,-

Vol. 341: Algebraic K-Theory I, Higher K-Theories. Edited by H. Bass. XV, 335 pages. 1973. DM 26,-

Vol. 342: Algebraic K-Theory II, "Classical" Algebraic K-Theory, and Connections with Arithmetic. Edited by H. Bass. XV, 527 pages. 1973. DM 36,-

Vol. 343: Algebraic K-Theory III, Hermitian K-Theory and Geometric Applications. Edited by H. Bass. XV, 572 pages. 1973. DM 38,-

Vol. 344: A. S. Troelstra (Editor), Metamathematical Investigation of Intuitionistic Arithmetic and Analysis. XVII, 485 pages. 1973. DM 34,-

Vol. 345: Proceedings of a Conference on Operator Theory. Edited by P. A. Fillmore. VI, 228 pages. 1973. DM 20,-

Vol. 346: Fučík et al., Spectral Analysis of Nonlinear Operators. II, 287 pages. 1973. DM 26,-

Vol. 347: J. M. Boardman and R. M. Vogt, Homotopy Invariant Algebraic Structures on Topological Spaces. X, 257 pages. 1973. DM 22,-

Vol. 348: A. M. Mathai and R. K. Saxena, Generalized Hypergeometric Functions with Applications in Statistics and Physical Sciences. VII, 314 pages. 1973. DM 26,-

Vol. 349: Modular Functions of One Variable II. Edited by W. Kuyk and P. Deligne. V, 598 pages. 1973. DM 38,-

Vol. 350: Modular Functions of One Variable III. Edited by W. Kuyk and J.-P. Serre. V, 350 pages. 1973. DM 26,-

Vol. 351: H. Tachikawa, Quasi-Frobenius Rings and Generalizations. XI, 172 pages. 1973. DM 18,-

Vol. 352: J. D. Fay, Theta Functions on Riemann Surfaces. V, 137 pages. 1973. DM 16,-

Vol. 353: Proceedings of the Conference. on Orders, Group Rings and Related Topics. Organized by J. S. Hsia, M. L. Madan and T. G. Ralley. X, 224 pages. 1973. DM 20,-

Vol. 354: K. J. Devlin, Aspects of Constructibility. XII, 240 pages. 1973. DM 22,-

Vol. 355: M. Sion, A Theory of Semigroup Valued Measures. V, 140 pages. 1973. DM 16,-

Vol. 356: W. L. J. van der Kallen, Infinitesimally Central-Extensions of Chevalley Groups. VII, 147 pages. 1973. DM 16,-

Vol. 357: W. Borho, P. Gabriel und R. Rentschler, Primideale in Einhüllenden auflösbarer Lie-Algebren. V, 182 Seiten. 1973. DM 18,-

Vol. 358: F. L. Williams, Tensor Products of Principal Series Representations. VI, 132 pages. 1973. DM 16,-

Vol. 359: U. Stammbach, Homology in Group Theory. VIII, 183 pages. 1973. DM 18,-

Vol. 360: W. J. Padgett and R. L. Taylor, Laws of Large Numbers for Normed Linear Spaces and Certain Fréchet Spaces. VI, 111 pages. 1973. DM 16,-

Vol. 361: J. W. Schutz, Foundations of Special Relativity: Kinematic Axioms for Minkowski Space Time. XX, 314 pages. 1973. DM 26,-

Vol. 362: Proceedings of the Conference on Numerical Solution of Ordinary Differential Equations. Edited by D. Bettis. VIII, 490 pages. 1974. DM 34,-

Vol. 363: Conference on the Numerical Solution of Differential Equations. Edited by G. A. Watson. IX, 221 pages. 1974. DM 20,-

Vol. 364: Proceedings on Infinite Dimensional Holomorphy. Edited by T. L. Hayden and T. J. Suffridge. VII, 212 pages. 1974. DM 20,-

Vol. 365: R. P. Gilbert, Constructive Methods for Elliptic Equations. VII, 397 pages. 1974. DM 26,-

Vol. 366: R. Steinberg, Conjugacy Classes in Algebraic Groups (Notes by V. V. Deodhar). VI, 159 pages. 1974. DM 18,-

Vol. 367: K. Langmann und W. Lütkebohmert, Cousinverteilungen und Fortsetzungssätze. VI, 151 Seiten. 1974. DM 16,-

Vol. 368: R. J. Milgram, Unstable Homotopy from the Stable Point of View. V, 109 pages. 1974. DM 16,-

Vol. 369: Victoria Symposium on Nonstandard Analysis. Edited by A. Hurd and P. Loeb. XVIII, 339 pages. 1974. DM 26,-

Vol. 370: B. Mazur and W. Messing, Universal Extensions and One Dimensional Crystalline Cohomology. VII, 134 pages. 1974. DM 16,-